ENVIRONMENTAL PLANNING FOR DESIGN AND CONSTRUCTION
    Martin N. Fabrick and Joseph J. O'Rourke
MANAGING CONSTRUCTION CONTRACTS: Operational Controls for Commercial Risks
    Robert D. Gilbreath
SUCCESSFUL METHODS IN COST ENGINEERING
    Hira N. Ahuja and Michael A. Walsh
QUANTITATIVE CONSTRUCTION MANAGEMENT: Uses of Linear Optimization
    Robert M. Stark and Robert H. Mayer, Jr.
ENERGY MANAGEMENT AND CONTROL SYSTEMS, Volume 1: Theory and Application
    Manuel C. Macedo, Jr.
PROJECT MANAGEMENT: Techniques in Planning and Controlling Construction Projects
    Hira N. Ahuja
COST ACCOUNTING FOR THE CONSTRUCTION FIRM
    Bill G. Eppes and Daniel E. Whiteman

COST ACCOUNTING FOR
THE CONSTRUCTION FIRM

COST ACCOUNTING FOR THE CONSTRUCTION FIRM

BILL G. EPPES
University of Florida
Gainesville, Florida

DANIEL E. WHITEMAN
Gulf Constructors International, Inc.
Tampa, Florida

A Wiley-Interscience Publication

JOHN WILEY & SONS

NEW YORK • CHICHESTER • BRISBANE • TORONTO • SINGAPORE

Copyright © 1984 by John Wiley & Sons, Inc.

All rights reserved. Published simultaneously in Canada.

Reproduction or translation of any part of this work beyond that permitted by Section 107 or 108 of the 1976 United States Copyright Act without the permission of the copyright owner is unlawful. Requests for permission or further information should be addressed to the Permissions Department, John Wiley & Sons, Inc.

Library of Congress Cataloging in Publication Data:
Eppes, Bill G.
  Cost accounting for the construction firm.

  (Construction management and engineering)
  Includes index.
  1. Construction industry—Accounting. 2. Cost accounting.　I. Whiteman, Daniel E.　II. Title.
III. Series.
HF5686.B7E66　　　1984　　　　657'.869042　　　　83-21752
ISBN 0-471-88537-1

Printed in the United States of America

10 9 8 7 6 5 4 3 2 1

SERIES PREFACE

Industry observers agree that most construction practitioners do not fully exploit the state of the art. We concur in this general observation. Further, we have acted by directing this series of works on Construction Management and Engineering to the continuing education and reference needs of today's practitioners.

Our design is inspired by the burgeoning technologies of systems engineering, modern management, information systems, and industrial engineering. We believe that the latest developments in these areas will serve to close the state of the art gap if they are astutely considered by management and knowledgeably applied in operations with personnel, equipment, and materials.

When considering the pressures and constraints of the world economic environment, we recognize an increasing trend toward large-scale operations and greater complexity in the construction product. To improve productivity and maintain acceptable performance standards, today's construction practitioner must broaden his concept of innovation and seek to achieve excellence through knowledgeable utilization of the resources. Therefore our focus is on skills and disciplines that support productivity, quality, and optimization in all aspects of the total facility acquisition process and at all levels of the management hierarchy.

We distinctly believe our perspective to be aligned with current trends and changes that portend the future of the construction industry. The books in this series should serve particularly well as textbooks at the graduate and senior undergraduate levels in a university construction curriculum or continuing education program.

JOHN F. PEEL BRAHTZ

La Jolla California
February 1977

PREFACE

The success of a general contracting company depends not only on quality construction but also on how well the company is able to control costs. Thin margins of profit, together with the numerous uncertainties involved in building construction, make it imperative to formulate a good system of cost control.

This book provides an effective system of cost accounting. This system presents a well-organized methodology for cost estimates, actual costs, cost forecasting, and reliable information for historical data.

The *cost estimate*, used to prepare a bid or negotiate a contract, serves as the initial instrument in a cost accounting system and is used to assure that all items of work have been reviewed.

All of the *actual costs* expended to construct a project must be recorded on a timely basis in a concise and well-ordered form. The cost accounting system used for each particular project must demonstrate how these actual costs to date relate to the estimated costs.

The cost accounting system provides a method to accomplish accurate *cost forecasting*. In the field of general contracting and construction management, the contracting organization is involved with very large sums of money, a very small percentage of which will become the contractor's fee for the project. For accurate information and assurance that a profitable operation is being conducted, a cost forecast must be projected throughout the course of the project, at any point in time during the project.

To establish probable future costs, a general contractor must rely on *historical data*—the contractor's past performance of similar work. This accumulation of data can establish a current (or future) cost estimate.

Information that serves as source material (input) for a cost accounting system must be accurate. One cannot expect to obtain reliable data on which to make value judgments, and one cannot expect these data to be any more accurate than the input. If costs are incorrectly distributed, they are of no value. In fact, they are quite misleading. Inaccurate data can cause errors which could prove very costly to a general contractor.

Everyone involved in the cost accounting system must be provided with written instructions explaining the system, and in particular the numerical breakdown which describes the activities and subactivities for each trade division. Reliable output is based on respect for the system and on accurate entries. The Glossary at the back of the book provides definitions of all the terms used throughout the Cost Coding System. Abbreviations of terms begin on page xvii.

Chapters 1–11 address the reporting and recording procedures. These chapters provide, in the chronological order of the project, detailed procedures used in the cost accounting of a project from beginning to end.

Chapter 12 contains the Cost Coding System, which is based on a 16-division format for organizing construction specifications. This format was developed by the Construction Specification Institute (CSI) and is published in the *CSI Manual of Practice*, MP-2-1, MASTERFORMAT-Master List of Section titles and numbers. This 16-division CSI format is used by most architects and engineers in the preparation of their specifications for a project.

For each division of the work there are subdivisions for cost codes and descriptions. The CSI format utilizes a five-digit system. The first two digits identify the major trade division (e.g., 01 for Division 1—General Conditions). The remaining three digits provide 999 possible subdivision codes within each division.

The Key Word Index at the back of the book is an alphabetical listing of the cost code terms that will assist in finding the appropriate cost code for various items of work.

Samples of each of the various forms, which are an integral part of this cost accounting system, are found in the appendix.

This book uses a general contracting company structure, as shown in Figure 1. Any contracting organization can identify its own structure, personnel, organization, and operations in accordance with the personnel in the chart and the description of the job and responsibilities listed below.

The president has overall responsibility for all of the operations of the company. This person usually signs contracts with owners and is responsible to the Board of Directors if the company is a corporation.

The division manager is assigned by the president to handle a specific group of projects. The president and division manager determine which projects the company will bid on or negotiate. The division manager assigns project managers to bid work and later to manage construction operations. The division manager supervises and works with his or her project managers during construction operations.

The project manager is responsible for estimating and bidding specific projects. The project manager, together with the division manager, assigns the superintendents to specific projects. This project manager has overall responsibility for the project and directs the activities of the superintendent. The project manager reviews the payroll and approves purchases and payments to subcontractors.

The superintendent is assigned to only one project at a time and is responsible for the day-to-day construction activities at the job site, supervising the activities of the various craft foremen, and scheduling and directing subcontractors' activities. The superintendent orders material to

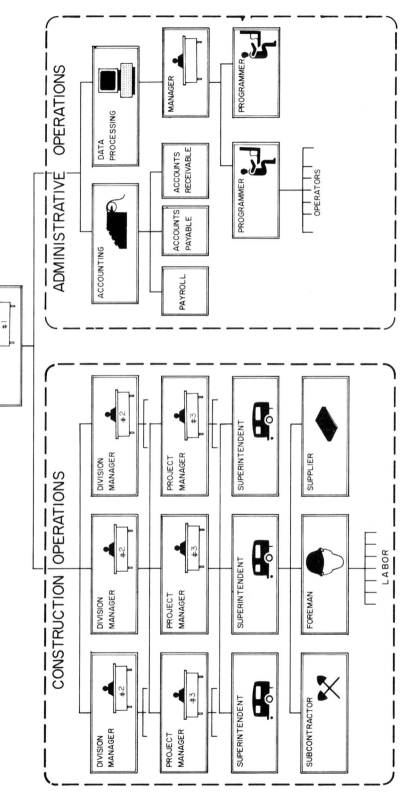

Figure 1. Organization chart of general contracting firm.

the job site that has been purchased by the project manager. The superintendent reviews time cards; completes the unit completion report for material quantities that have been placed; handles applications for employment, employee tax forms, and shipping invoices; and forwards these to the project manager.

The foreman directs the activities of the construction workers, prepares time cards, and enters cost code numbers on the time cards.

Accounting receives and codes various requests involving payment after they have been approved by the project manager. These include time cards, purchases, and subcontractor request for payment. In addition, Accounting is responsible for administering all overhead costs.

Data Processing receives its documents from Accounting and enters this information into the computer, where it is stored for future use.

The Cost Accounting System presented in this book addresses two basic areas for which accounting is vital:

1. Invoices
   a. Subcontractor application for payment
   b. Purchase orders
   c. Material (no purchase order issued)
2. Payroll for labor

The reporting and recording procedures in this book are presented in relative chronological order as they relate to each individual project. This will provide a uniform method to demonstrate the use of the Cost Accounting System by all personnel involved in the course of project accountability.

Individual sections may overlap in the discussion. This is not intended to be repetitious, but rather to demonstrate the interrelationships of the various sections involved in these procedures.

Most of the sections consist of two basic functions. These functions are often referred to as input and output. The input function is performed by the project manager in furnishing data to Accounting and Data Processing for implementing the various systems. The output function consists of the resultant forms and reports which are furnished by Accounting and Data Processing for use by the project manager in keeping track of costs related to the project. Output is also used to provide cost data for future negotiation and bid work.

BILL G. EPPES
DANIEL E. WHITEMAN

Gainesville, Florida
Tampa, Florida
February 1984

CONTENTS

List of Figures, xv
Abbreviations, xvii

CHAPTER 1 ESTIMATE LISTING   1

CHAPTER 2 SOURCE DOCUMENT INPUT   7

CHAPTER 3 INVOICE APPROVAL AND INVOICE QUANTITY INPUT   12

  Description/Responsibility of Invoice Approval Form, 12

CHAPTER 4 LABOR COSTS AND IN-PLACE QUANTITY INPUT   19

  Weekly Time Cards, 19
  Unit Completion Report, 20

CHAPTER 5 SUMMARY REPORTS   23

  Subcontract, Purchase Order, and Materials Summary, 24
  Labor Summary, 30

CHAPTER 6 VARIANCE REPORTING PROCEDURES   33

  Subcontract, Purchase Order and Material Summary, 34
  Labor Summary, 35

CHAPTER 7 PROJECT STATUS REPORT   37

Project Structure, 37
Billing Structure, 39
Budget Analysis, 41
Completion Status, 41
Owner's Representative Status, 42
Project Manager Remarks, 42
Division Manager Remarks, 42

CHAPTER 8 MAINTENANCE OF ESTIMATED QUANTITIES AND COSTS   43

Input of Projected Final Quantities 43
Change Orders, 44
Net Redistribution of Estimated Costs, 48

CHAPTER 9 MAINTENANCE OF ACTUAL COSTS AND QUANTITIES   53

CHAPTER 10 ACCUMULATION OF EXTRAORDINARY COSTS   57

Subcontractor/Vendor Backcharges, 57
Owner/Architect/Engineer Force Accounts, 58
Major Corrective Work, 59

CHAPTER 11 PROJECT CLOSE-OUT PROCEDURES   60

Accounting Close-out Procedures, 60
Historical Data Base, 66
Postwarranty Close-out Procedure, 67

CHAPTER 12 COST CODING SYSTEM   68

Division 1—General Conditions, 68
Division 2—Site Work, 77
Division 3—Concrete, 83
Division 4—Masonry, 93
Division 5—Metals, 94
Division 6—Woods and Plastics, 96

Division 7—Thermal and Moisture Protection, 100
Division 8—Doors and Windows, 102
Division 9—Finishes, 104
Division 10—Specialties, 106
Division 11—Equipment, 108
Division 12—Furnishings, 120
Division 13—Special Construction, 121
Division 14—Conveying Systems, 125
Division 15—Mechanical, 127
Division 16—Electrical, 139

GLOSSARY 140

APPENDIX: COST ACCOUNTING FORMS 147

KEY WORD INDEX 173

LIST OF FIGURES

1.	Organization Chart of General Contracting Firm	ix
1.1.	Accounting and Construction Management Flow Chart	2
1.2.	Estimate Listing Form (input source)	3
1.3.	Estimate Listing Form (output)	5
2.1.	Source Document Input	8
3.1.	Invoice Approval Form	13
3.2.	Invoice Approval Flow Chart	15
3.3.	Subcontractor's Application for Payment	17
3.4.	Accounting Error Review Process	18
4.1.	Payroll Flow Chart	20
4.2.	Time Card Form	21
4.3.	Unit Completion Report (printout)	21
5.1.	Subcontract, Purchase Order and Material Summary (printout)	24
5.2.	Monthly Transaction Report (printout)	29
5.3.	Weekly Labor Summary	31
5.4.	Monthly Labor Summary (printout)	31
7.1.	Project Status Report	38
8.1.	Maintenance Update—Estimates (printout)	44
8.2.	Estimate Listing and Maintenance Form (blank form)	45
8.3.	Maintenance Form 1	47
8.4.	Maintenance Form 2	49
8.5.	Maintenance Form 3	50
8.6.	Maintenance Form 4	51
9.1.	Actual Cost and Quantity Maintenance Form	54
11.1.	Commitment Run (printout)	61
11.2.	Project Close-Out Report	64

ABBREVIATIONS

AC	Acre
AVG	Average
BF	Board feet
CC	Cost code
CF	Cubic feet
CUM	Cumulative
CY	Cubic yard
EA	Each
EST	Estimate
HP	Horsepower
HRS	Hours
HVAC	Heating, ventilating, and air conditioning
ID	Inch diameter
LF	Linear feet
LH	Labor hour
LH/U	Labor hours per unit of measure
LS	Lump sum
MAT	Material
MO	Month
N/C	No charge
NO	Number
PC	Piece
P&L	Profit and loss
PO	Purchase order
PROD	Production
QTY	Quantity
RETAIN	Retainage
SF	Square feet
SQ	Square; equals 100 square feet
SUB	Subcontract

SY	Square yard
TN	Ton
U/LH	Units per labor hour
UM	Unit of measure
WIP	Work in process
WKLY	Weekly
YR	Year

CHAPTER 1

ESTIMATE LISTING

One of the first functions performed by a project manager on a newly acquired project is to provide Accounting with the input data required to account for the costs of the project as estimated. A flow chart illustrating the interactions among the various areas of Accounting and Project Management is provided in Figure 1.1. It explains the responsibilities and duties of key personnel and/or departments within a construction organization.

These data are submitted to Accounting on the Estimate Listing and Maintenance Form shown in Figure 1.2. This form serves a dual purpose. When it is used for estimate listing, the word *maintenance* is crossed out of the title on the form. Uses for the maintenance form will be discussed in Chapter 8.

The information to be put into the estimate listing is the cost "estimate" of the project. This is the original estimated cost of the total project. The estimate listing must total the exact cost of the estimate. The difference between contract amount and cost will then represent the gross profit on the project.

Column 1 on the estimate listing is for the cost code, which is obtained from the Cost Coding System in Chapter 12 and should be used as a checklist in the preparation of the estimate listing. As shown in Figure 1.2, there is a cost code entry #03322. The number 03, taken from the Cost Coding System in Chapter 12, refers to the construction code for Division 3, which is concrete. The last three numbers of the cost code (322) refer to a subitem in the major division of concrete, in this case slabs on grade. These printed cost codes are the only permanent cost codes that are included within the system. Upon approval, the permanent cost codes may be modified to a nonstandard cost code for a particular project.

Column 2, document type and number, contains four basic types of documents:

1. Subcontract
2. Purchase order
3. Material
4. Labor

The various types of documents should be in the order listed above. The document type and number of subcontracts and purchase orders in this column will take one of two forms: first, when a source document has been prepared; and second, when the source document may not have been prepared at the time of the estimate listing.

In the first case, the document type and the document number should be included. For example, subcontract agreement 0627 would be shown on the form as S-0627, and purchase order 2890 would show the document type and number as P-2890.

In the second instance, as in the case shown in Figure 1.2, the document type and number for subcontracts and purchase orders should be shown only as SUB-1, OR PO-1, respectively. If more than one subcontract or purchase order are anticipated in a cost code, the estimated cost for each is

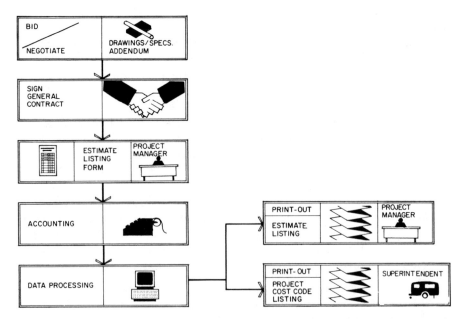

Figure 1.1. Accounting and construction management flow chart. As soon as the contract is signed, the project manager provides Accounting with the necessary data to initiate the new project into the Cost Accounting System.

ESTIMATE LISTING 3

ESTIMATE LISTING & MAINTENANCE FORM

NAME SAMPLE   JOB NO. XXXX   MAINTENANCE FORM NO._____
                                                          CHANGE ORDER NO._____

COST CODE	DOCUMENT TYPE #	ESTIMATED QUANTITY	U/M	ESTIMATED TOTAL MH	ESTIMATED TOTAL COST	PROJECTED FINAL QUANTITY
03322	SUB-1	200			600	
03322	MAT	200			10,000	
01802	LABOR	200		25	100	
01802	LABOR				20	
TOTALS					10,720	

Submitted by:_____XXXX_____ Date:_____
Processed by (Accounting):_____ Date:_____   PAGE____OF____
Processed by (Data Process.):_____ Date:_____

Figure 1.2. Estimate listing form (input source). The project manager provides project information to Accounting, which is important for effective cost management of the project.

to be shown as SUB-1, SUB-2, and so on, or PO-1, PO-2, and so on, on the estimate listing. In the cases where SUB-1 is used, the change to the subcontract number must be shown on the upper right-hand corner of the document.

Occasionally a subcontractor will perform work without a subcontract agreement. In these instances, the estimate listing should be completed with only the term SUB shown. This will indicate to Accounting that no subcontract agreement is forthcoming.

The cost for items of material will be shown in the cost code as only one total. They cannot be broken down into each individual item, as can subcontracts and purchase orders. If this type of control is required within the cost code, it may be obtained by writing purchase orders for the separate units of material.

The last document type appearing in this column is all cost for labor within the cost code.

Column 3 is used to denote the quantity of an item estimated. If an item is lump sum, no quantity has to be entered in this column. *Variance Reporting Procedures* (discussed in a Chapter 6) explain the input required to report variance on lump sum items that are not covered by firm source documents.

Column 4 is used whenever *alternate* units of measure are used within a cost code. The intended unit of measure is shown in the Cost Coding System immediately following the cost code description. Very often in the case of subcontracts, the project manager may decide to change from a unit of measure in a particular cost code to a lump sum contract. In these instances, the unit of measure column is to be filled in with the notation LS to indicate that the unit of measure has been changed to lump sum. This column is to be used only when there is a change in units of measure. It is not to be used when the unit of measure remains the same as shown in the Cost Coding System.

Column 5 is to be used *only* in those items involving labor. This column is the input for estimated total labor hours in the cost code.

Column 6 is the estimated total cost of each type of cost within the cost code. This is the cost for each type of subcontract, purchase order, material, and labor. This cost code is not to be totaled. The total for each cost code will be done automatically in Data Processing.

The estimated total cost column should be totaled to assure that the estimated total cost of the estimate agrees with the total cost shown on the estimate listing.

Column 7 is to be used for projected final quantity when the quantity has been proofed by the project manager at the time of preparation of the estimate listing. If the quantity is not proofed until later, this column is to remain blank until that time.

Columns 2 through 6 in Figure 1.2 also show the following:

1. *Subcontract.* This is a subcontract for the placement of 200 cubic yards of concrete. The subcontract amounts to $3 per cubic yard for a total cost of $600.

2. *Material.* The cost for concrete amounts to 200 cubic yards at $50 per cubic yard (including tax) for a total cost of $10,000.

```
ESTIMATE LISTING                                                                                           PAGE 00
DIV. _____        PROJ. MGR. _____              JOB NO. & NAME 0000    DATE RUN 0/00/00
                                                                         SAMPLE JOB
*********************************************************************************************************
COST CODE           UM  *                     C U R R E N T   E S T I M A T E   D A T A
- - - - - - - - - - -  *
*********************************************************************************************************
                                                                                    EST                 ESTIMATED    PRODUCTIVITY
CODE CODE DESCRIPTION        EST. QTY.   EST MAN HRS   EST TOT COST   PROJ. QTY.   EST UNIT COST   AVG WG      U/MH           MH/U
00000                   UM
SAMPLE COST CODE
  SUB.                           .00                        .00                         .00
  P.O.                           .00                        .00                         .00
  MAT.                           .00                        .00                         .00
  LAB.                           .00         .00            .00                         .00            .00         .00            .00

COST CODE TOTAL                                             .00
*********************************************************************************************************

SUB. JOB ESTIMATE TOTALS      0.00                         0.00

P.O. JOB ESTIMATE TOTALS      0.00                         0.00

MAT. JOB ESTIMATE TOTALS      0.00                         0.00

LAB. JOB ESTIMATE TOTALS      0.00         0.00            0.00

TOTAL JOB ESTIMATE FOR
  SAMPLE JOB                  0.00         0.00            0.00
```

Figure 1.3. Estimate listing form (output). After Accounting receives the project input from the project manager, Data Processing produces a printout of this information.

3. *Labor.* Per agreement, the company is to furnish a chutetender during placement of the concrete. It has been established that 25 labor hours at $4 per hour will be required to place the 200 cubic yards for a total cost of $100. The attached estimate listing shows the input data required to complete this form.

4. *Labor Tax.* Whenever labor is expended on a project, a proportionate cost must be added to the estimate. Cost Code 01802 is used for this cost, which, for example, at 20% of the labor cost ($100), amounts to $20.

The completed estimate listing is entered into the Cost Accounting System for processing. It may not always be possible to complete the estimate listing prior to the start of construction on the project. Therefore, Accounting may open up a few cost codes at the beginning of the project by notification from the project manager. However, this practice should be kept to a minimum and the estimate listing completed as soon as possible. The accounting supervisor will then have the estimate listing form keypunched and printed by Data Processing. The completed printout of the estimate listing, when keypunched and printed for the project manager, is shown in Figure 1.3. This printout will show each of the cost codes broken down into the various types of cost. The printout will be totaled at the bottom of the report for each of the types of cost, and there will also be a grand total showing the total cost of the estimate. These totals must be checked against the estimate to assure accuracy.

The second output to the project manager is the project cost code listing. As explained under *Net Redistribution of Estimated Costs* in Chapter 8, the superintendent is to be given only the cost codes and descriptions applicable to his or her project. To avoid needless typing, a computer program has been established to reproduce the cost codes and descriptions as they appear in the Cost Accounting System. The cost codes and descriptions applicable for each project will be printed directly from the estimate listing completed by the project manager. The system for each project will be printed in duplicate on the same size and type cards as the weekly time cards (see Figure 4.3 in Chapter 4). Copies of the system for that project are to be distributed to the project manager and superintendent.

As the estimate listing is the original estimate, it is to be *input only once* on each project. Any subsequent maintenance to the estimate, such as owner change orders, projected final quantities, or redistribution of costs, is to be entered through the use of the maintenance form, as discussed in Chapter 8.

CHAPTER 2

SOURCE DOCUMENT INPUT

The two types of originating documents are subcontracts and purchase orders. In order for subcontracts and purchase orders to become a part of the Cost Accounting System, certain information must be placed on the source documents submitted for entry into the system. This chapter deals with the items required to be placed on a source document for input into the system. (See the appendix for sample of purchase order and subcontract agreement, which shows preprinted number and space for input data.) A source document flow chart is shown in Figure 2.1.

There are two guidelines to remember when preparing subcontracts and purchase orders for input into the accounting system.

1. Subcontracts and purchase orders *may not* apply to more than one project number.

2. Subcontracts and purchase orders *may* apply to more than one cost code within a single project.

Certain information must be shown in the upper right-hand corner of the source document below the document number. The order in which the information is to appear on the document is as follows:

1. Directly below the document number put the project number.

As an example:

Subcontract Number
Show this number on invoices
No. 0799
Project Number: 5060

2. Directly below the project number put the appropriate cost code(s) to which the source document applies. If more than one cost code is involved, list one beside the other.

In this example, subcontract 0799 is assigned to Cost Code 09300, which, referring to Division 9 in Chapter 12, is hard tile. Cost Code 10270 is

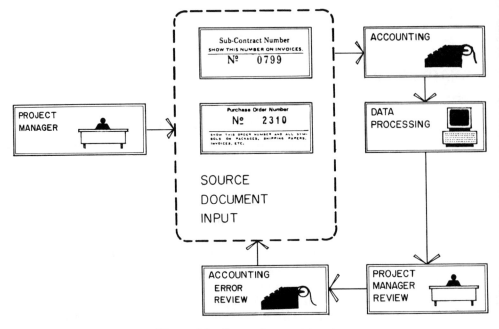

Figure 2.1. Source document input.

As an example:

Subcontract Number
Show this number on invoices
No. 0799
Project Number: 5060
Cost Code: 09300   Cost Code: 10270

access flooring. These items of work will be supplied by the same subcontractor.

3. Directly below the cost code(s) show the document type and number shown on the estimate listing. If the source document has been written prior to the preparation of the estimate listing, the source document type and number should be as shown on the document. However, if the estimate listing preceded the source document, it will be necessary to substitute the document type and number for the SUB-1 or PO-1 that appeared on the estimate listing.

The first item shown would be the document type and number as shown on the estimate listing. Directly next to this is to be shown the document type and number replacing that of the estimate listing.

SOURCE DOCUMENT INPUT

As an example:

Subcontract Number
Show this number on invoices
No. 0799
Project Number: 5060

Cost Code: 09300   Cost Code: 10270
SUB-1 = S0799   SUB - 1 = S0799

It should be noted that the subcontracts shown above were the first subcontracts issued in the respective cost code numbers.

4. Directly below the document type and number state whether the document is to be *lump sum* or *unit price*. This item determines the proper variance procedures to be followed by the computer. The variance procedures for source documents are discussed further in Chapter 6.

It is possible to write lump sum subcontracts and purchase orders for any cost code. This applies to all cost codes including those with other units of measure. The variance reporting procedures followed by Data Processing for lump sum documents will be implemented for any cost code for which the source document input is noted as lump sum.

5. At this point, the additional information to be entered on the source document depends on whether the source document is lump sum or unit price. These two items are discussed separately below:

a). Lump sum documents. If the document has been written for a lump sum amount, simply put the document amount to be paid directly below the lump sum.

As an example:

Subcontract Number
Show this number on invoices
No. 0799
Project Number: 5060

Cost Code: 09300   Cost Code: 10270
SUB-1 = S0799   SUB-1 = S0799
Lump Sum   Lump Sum
$10,000 $5,000

There may be times when the project manager determines that it is not necessary to maintain the cost records separately for the different cost codes. In that instance, it is possible to combine the costs for the total subcontract into the lowest cost code number for the work included. However, for historical data purposes, the project manager should maintain the separate costs for later use.

b). Unit Price documents. If the document has been written for a unit price, the next information that is required is the quantity and unit measure, followed by the unit price amount of the source document.

Multiple cost codes and unit prices may be used on a purchase order; however, only one unit price per cost code may be entered on each purchase order. The accounting system cannot accept a purchase order that has a cost code listed with more than one unit price.

As an example:

Subcontract Number
Show this number on invoices
No. 0799
Project Number: 5060

Cost Code: 09300	Cost Code: 10270
SUB-1 = S0799	SUB-1 = S0799
Unit Price	Unit Price
Quantity = 8,000 SF	Quantity = 2,000 SF
Unit Price = $1.25	Unit Price = $2.50

6. Many times a project manager will write one subcontract or purchase order for numerous small items of materials in order to protect the quoted price. In these instances, there is often no need for the accounting system to control this document or print variances for these items.

In these cases, the project manager is to write in the space below the source document the term "For the file only," followed by the project number and then by the cost codes to which the subcontract or purchase order applies. The last item is the document type and number shown on the estimate listing, along with the actual number of the source document preceded by the letters FP (file purchase order only), or FS (file subcontract only). Accounting will then keep the purchase order as a file copy only, and the purchase order number will then show in each cost code applicable. In these instances, the items included in this purchase order should still be coded as a purchase order item.

As an example:

Purchase Order Number
No. 2310
Show this order number and all
symbols on packages, shipping
papers, invoices, etc.

Contract _____
Location _____
Date   _____
Terms   _____

For file only
Project Number: 5060
Cost Codes: 03022/03325/03326
PO-1 = FP2310

The above information is to be typed on all copies of the source document. The subcontractor should then include the project number, subcontract and/or purchase order number, and cost code(s) on the subcontractor's application for payment form or supplier's invoice when requesting payment. This is further covered in Chapter 3.

CHAPTER 3

INVOICE APPROVAL AND INVOICE QUANTITY INPUT

The processing of *accounts payable* constitutes a major portion of the work in process on any given project. Accounts payable consists of three basic items:

1. Subcontracts
2. Purchase orders
3. Material (defined as a purchase where no purchase order is issued)

These accounts payable are processed through the Cost Accounting System by use of an Invoice Approval Form.

The invoice approval form, shown in Figure 3.1, is used to input into the accounting system two essential items, invoice cost and invoiced quantities.

Each space of the invoice approval form serves a designated purpose. The procedure for filling out this form is described below. The area of responsibility for completion in each of the spaces is illustrated in Figure 3.1.

DESCRIPTION/RESPONSIBILITY OF INVOICE APPROVAL FORM

1. *Mo.–Yr.* The month and year in which the invoice is chargeable to work in process. Responsibility of Accounting.

2. *Vendor No.* The number that is established by Accounting for each subcontractor and supplier. Responsibility of Accounting.

3. *P.O. No.* The purchase order number shown on the source document issued to the supplier. Responsibility of Accounting.

4. *Blank.* Check (x) signifies the invoice is to be charged as a purchase order item. Responsibility of the project manager.

Checked (x) for all purchase order items, including those purchase

DESCRIPTION/RESPONSIBILITY OF INVOICE APPROVAL FORM

INVOICE APPROVAL FORM

MO. – YR. (1)	VENDOR NO. (2)	P.O. NO. (3)	(4)	SUB-CONTRACT NO. (5)	(6)
	JOB # (7)		MAT'L. (8)	SUB (9)	
DUE DATE (10)	INVOICE # (11)	COST CODE (12)	QTY. (13)	$ AMOUNT (14) .00	CODE (15)
APPROVED BY (16)					

Figure 3.1. Invoice approval form.

orders that have been processed "For file only." Refer to Chapter 2, Source Document Input, for further explanation.

5. *Subcontract No.* Provided for the subcontract number shown on the source document issued to the subcontractor. Responsibility of Accounting.

6. *Blank.* Checked (x) to signify that the invoice is to be charged as a subcontract item. Responsibility of the project manager.

7. *Job #.* The project number to which the vendor's invoice is to be charged. Only one project number may be used per the invoice approval form. If vendor bills for more than one project, additional invoice approval forms must be used for each project. Responsibility of the project manager.

8. *Material.* Checked (x) if the invoice is for material only. Responsibility of the project manager.

9. *Sub.* Check (x) if the invoice has been issued for subcontract agreement. This space is *not* to be used if a subcontract agreement has been issued that covers this invoice. Responsibility of the project manager.

10. *Due Date.* The date when payment for this invoice is due. Payments should be made only per the contract agreements. Processing of payments between contract dates creates obvious negative cash flow within the corporation. Payment exceptions must be held to the essential minimum and require approval of the president. Responsibility of the project manager.

NOTE: The project manager is responsible for establishing the payment date. This includes responsibility for seeing that the payment date coincides with the supplier's discount date. Accounting is to take all discounts regardless of payment date. If payment date and discount date do not coincide problems will inevitably occur.

11. *Invoice #.* The invoice number that appears on the billing may be shown on the check to the firm. Six separate invoices can be processed on this form, but they must be applicable to the same project and firm. Responsibility of the project manager.

12. *Cost code.* The cost code to which the invoice is applicable. Responsibility of the project manager.

NOTE: If there are several cost codes applicable to a single invoice, the project manager is to show directly on the invoice the applicable cost code, gross amount, quantity, and unit of measure to be charged to each code. Accounting will then transfer this information to the invoice approval form.

13. *Qty.* The number of units being paid for under this invoice. This quantity may be shown to two decimal places if this accuracy is desired (7.5 CY or 1.55 EA). Responsibility of the project manager.

NOTE: This space may also be used to represent the percent of total cost that this invoice represents, as explained in Chapter 1, Estimate Input, and Chapter 6, Variance Reporting Procedures. The percent complete shown in this space is to be only the percent of cost that this invoice represents.

14. *Amount.* The amount to be paid on the invoice. Responsibility of both the project manager and Accounting. The project manager is responsible for circling the amount to be paid directly on the invoice, and verifying the quantities, units of measure, unit prices for materials, and unit prices on purchase order. Accounting is responsible for all extensions, the total amount of the invoice, application of discounts and sales tax, and the transfer of all information in the amount column to the invoice approval form. Accounting is also responsible for seeing that the total amount of lump sum purchase orders and subcontracts is not exceeded.

15. *Code.* A code used solely by Accounting to signify the type of cost to which the invoice is applicable. The code references used are as follows: S = Subcontract, P = Purchase Order, R = Retainage, and M = Material. Responsibility of Accounting.

16. *Approved by.* To be signed by the individual authorized to approve the invoice. Responsibility of authorized personnel.

NOTE: The information shown on the invoice approval form will be keypunched by Data Processing directly from this form. Therefore, it is essential that all information be plainly printed.

The invoice approval flow chart involving the invoice approval form is shown in Figure 3.2. As invoices are received, they will be distributed

DESCRIPTION/RESPONSIBILITY OF INVOICE APPROVAL FORM

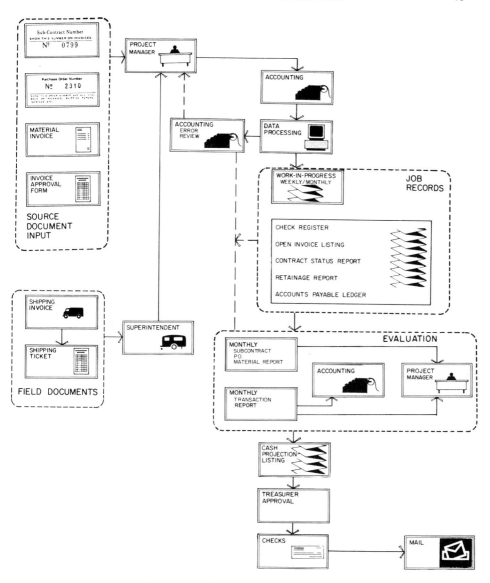

Figure 3.2. Invoice approval flow chart.

directly to the project manager. Invoices are to be promptly processed by the project manager or returned to the vendor. It should be the policy of a company not to make corrections to vendors' invoices. An incorrect invoice is to be returned promptly to the vendor for resubmission. For this reason, it is essential that invoices be corrected and resubmitted by the vendor without being delayed to the next pay period. Accounting will collect all invoices on a daily basis, and process through the accounting system immediately. The vendor's check will not be mailed until the due date filled in by the project manager.

All subcontractors are to invoice on the Subcontractor's Application for Payment Form provided by the company. A copy of this Subcontractor's application for payment is shown in Figure 3.3. The application for payment is to be completely filled out, including notarized lien releases.

After processing by Data Processing, the invoice will be included in the appropriate area of the next monthly summary report. This is further defined and discussed in Chapter 5, Summary Reports.

Invoices for subcontract agreements and purchase orders may be processed and paid only in the name of the firm indicated on the subcontract or purchase order agreements. Certain documentation must be provided Accounting to process any invoice from a firm name other than shown on the subcontract agreement or purchase order. There are two cases in which this situation might occur:

1. Contracts assigned by the subcontractor that the company has approved
2. An invoice being mailed directly from a manufacturer for purchases through a jobber

Assigned contracts are to be furnished to Accounting with the assignment agreement executed by the company after the president's approval.

Direct mail invoices, if it is known in advance that the manufacturer is going to bill direct, should have the purchase order made out to the manufacturing company rather than the jobber. If this is not done, written authorization must be received from the jobber to pay directly to the manufacturer. Accounting should be given a copy of this documentation.

As mentioned above, Accounting has the responsibility to see that lump sum subcontract and purchase order agreements are not overpaid. To ensure against the possibility of mistaken overpayment on these source documents, Accounting has been authorized to accept *no verbal approvals* for overpayment of the amount shown on lump sum subcontracts and purchase orders.

If a situation arises wherein the project manager agrees to pay an additional amount of a document, then a change order must be prepared, and Accounting must be given their copy of the change order.

It often happens that the invoice is received in excess of the source document amount when the document is written without either sales tax or freight included. The invoice from the supplier most often arrives with these items included. For example, the project manager writes a purchase order for $1,200 plus sales tax and freight. The invoice arrives in the amount of $1,280, which includes $64 for sales tax and $16 for freight.

One solution is to have all purchase orders and subcontracts written to include all applicable sales tax and freight shown separately on the document. If this is not done, the project manager must prepare a change order to increase the purchase order amount to $1,280 and give Accounting a copy of this change order.

If an invoice approved for payment exceeds the source document amount,

DESCRIPTION/RESPONSIBILITY OF INVOICE APPROVAL FORM

SUBCONTRACTOR'S APPLICATION FOR PAYMENT

FROM: _____
_____

PROJECT:
PAYMENT REQUEST NO. _____
PERIOD _____, 19___, to _____, 19___.

STATEMENT OF CONTRACT ACCOUNT:

1. Original Contract Amount — $ _____
2. Approved Changes (Net) (Add/Deduct) (As per attached breakdown) — $ _____
3. Adjusted Contract Amount — $ _____

4. Value of Work Completed to Date: (As per attached breakdown) — $ _____
5. Value of Approved Change Orders Completed to Date:
   (As per attached breakdown) — $ _____
6. Materials Stored on Site: (As per attached breakdown) — $ _____
7. Total (4 + 5 + 6) — $ _____
8. Less Amount Retained (_____ %) — ($ _____)
9. Total Less Retainage — $ _____
10. Less Previous Payments — ($ _____)
11. AMOUNT OF THIS REQUEST — $ _____

CERTIFICATE OF THE SUBCONTRACTOR:

   I hereby certify that the work performed and the materials supplied to date, as shown on the above represent the actual value of accomplishment under the terms of the Contract (and all authorized changes thereto) between the undersigned and _____ relating to the above referenced project.

   I also certify that all laborers, materialmen, suppliers, contractors, and subcontractors used on or in connection with the performance of this contract have been paid in full, except as noted on reverse side. I further certify I have complied with all Federal, State and local tax laws, including Social Security laws and Unemployment Compensation laws and Workers' Compensation laws insofar as applicable to the performance of this Contract.

   Furthermore, in consideration of the payments received, and upon receipt of the amount of this request, the undersigned does hereby waive, release and relinquish any and all claims under any applicable surety bond, rights of lien upon the above premises, and causes of action which the undersigned may now have or hereafter acquire, including, but not limited to those rights as contemplated by Chapters 255 and 713, Florida Statutes, except for rights to the extent that payment is retained pursuant to written agreement or payment to become due for work performed subsequent to the date hereof.

Date _____

Subscribed and sworn before me this _____ day of
_____, 19 _____

Notary Public: _____
My Commission Expires:

SUBCONTRACTOR
_____
BY: _____
         (authorized signature)

TITLE: _____

TWO PART: WHITE COPY — MAIL TO   YELLOW COPY — SUBCONTRACTOR'S COPY

Figure 3.3. Subcontractor's application for payment.

it will be returned to the project manager for either correction or adjustment to the source document and payment will fall into the next scheduled period.

Quantity is involved in every operation of a construction project. In this chapter we have referred to *invoiced quantity*. This will be shown on the subcontract, purchase order, and material monthly summary. The sum-

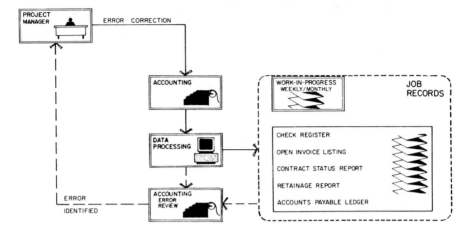

Figure 3.4. Accounting error review process.

mary report will show the invoiced quantity on a monthly basis and also the cumulative quantity invoiced to date.

The second type of quantity is *in-place quantity*. This quantity applies directly to the labor expended to install the materials. This quantity will be reported on the labor summary and will be discussed further in Chapter 4, Labor Cost and In-Place Quantity Input.

Occasionally a project manager will approve an invoice to a cost code that is not valid for the project. Cost code 99997 has been established as a temporary account for these "cost code errors." Without this temporary cost code the entire accounting process would be stopped until the correct cost code was obtained. This account is to be cleared up immediately by the project manager upon receipt of the summary indicating this cost code.

A portion of the invoice approval flow chart that illustrates the accounting error review process is shown in Figure 3.4.

Cost coding errors and other adjustments required to correct and update actual cost and invoiced quantity are accomplished through the use of the Actual Cost and Quantity Maintenance Form (see Figure 9.1) discussed in Chapter 9, Maintenance of Actual Costs and Quantities.

CHAPTER 4

LABOR COSTS AND IN-PLACE QUANTITY INPUT

As mentioned in the preface, the second major area in the Cost Accounting System is labor. This chapter discusses the procedure for providing input of labor costs into the Cost Accounting System, along with the input of the quantities that have been placed by that labor. Figure 4.1, a Payroll Flow Chart, illustrates this process.

Labor cost is, of course, made up of hours worked and average wage. In order accurately to forecast the projected final cost in each cost code, a production rate based on in-place quantity is required.

For this reason, the quantities applicable to labor do not necessarily correspond to the quantities that have been invoiced to date. With the exception of concrete, very few items are installed immediately upon invoice.

The output from this labor data consists of the weekly and monthly labor summary reports discussed in further detail in Chapter 5, Summary Reports. Other labor outputs furnished are the preprinted *weekly time cards* and *unit completion reports*.

WEEKLY TIME CARDS

The cost of labor is determined from labor information obtained from the field and furnished to the Cost Accounting System on weekly time cards, shown in Figure 4.2.

Each new employee completes the employment application and employee W-4 forms. These forms should be *completely* filled in by the employee, with the exception of the pay rate and approval, which are completed by the hiring supervisor.

The employment application is reviewed by the hiring supervisor who determines the employee's correct classification. The employment application and W-4 forms must then be forwarded to Accounting with the next scheduled payroll in order to assure timely processing of output data.

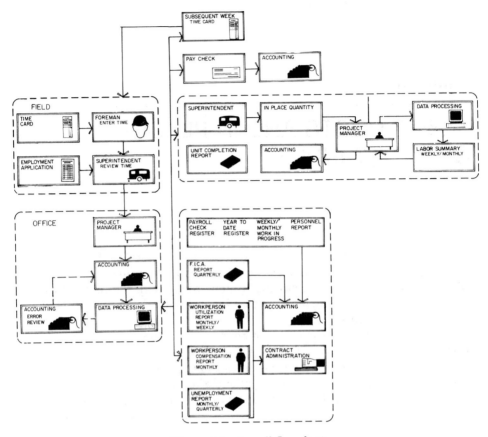

Figure 4.1. Payroll flow chart.

Accounting will print a weekly time card for that employee until the employee's status on the particular project has changed.

The hours worked by the employee should be filled out on the weekly time card furnished with the weekly time package. This time card is completed by the field superintendent and forwarded directly to the project manager for review and transmittal to the payroll clerk in Accounting.

UNIT COMPLETION REPORT

The Unit Completion Report, in Figure 4.3, has been developed for the purpose of inputting in-place quantities to date. The input of the in-place quantities must originate in the field. Therefore, the field superintendent is to report the installed quantities for all items of labor that are performed during the week corresponding to the weekly time cards. A quantity is to

TIME CARD

JOB # ________ JOB NAME ______________________________ WEEK END ________

EMP # ________ EMP NAME ______________________________ CRAFT CODE ________

DATE →																	
DAY OF WEEK →																	
C.C.#	ALT. CRAFT CODE	TOTAL STRAIGHT	TOTAL OT	STRAIGHT	OT	STRAIGHT	OT	STRAIGHT	OT	STRAIGHT	OT	STRAIGHT	OT	STRAIGHT	OT	STRAIGHT	OT
TOTAL																	

Figure 4.2. Time card form.

```
                              UNIT COMPLETION REPORT

 JOB #0000        JOB NAME      SAMPLE JOB                    WEEK ENDING  0/00/00
**************************************************************************************
*                                          *    *             * PREVIOUS    *           *           *
* COST     COST CODE DESCRIPTION           *UM  * ESTIMATED   * TOTAL QTY   * QUANTITY  *  TOTAL    *
* CODE                                     *    * QUANTITY    *   AS OF     * COMPLETE  * QUANTITY  *
*   #                                      *    *             *  0/00/00    * THIS WK   * COMPLETE  *
*                                          *    *             *             *           *           *
**************************************************************************************
*                                          *    *             *             *           *           *
* 00000    SAMPLE COST CODE                *    *     0.00    *      0.00   *    0.00   *    0.00   *
*------------------------------------------*----*-------------*-------------*-----------*-----------*
```

Figure 4.3. Unit completion report (printout).

be input for all cost codes having a labor charge for that period. It must be assumed that if labor has been charged to a cost code, then some indentifiable quantity has been installed. The quantity to be reported on lump sum cost codes is to be recorded as a percentage; all other cost codes with units of measure are to be reported by that applicable unit of measure.

The unit completion report form will be sent weekly to the field superintendent with the payroll package. This form will provide the field with a preprinted listing of each cost code for the particular project, and will include the estimated quantity for each cost code, the unit of measure for each cost code, and the quantity that has been reported as complete through the ending date shown on the report.

The superintendent will complete the report *each week* by inserting the quantity completed that week on all active items. The superintendent may also fill in the last column, quantity completed to date, if he or she so desires. This is not mandatory as it will be done automatically by the computer. When completed, the unit completion report is to be returned with the weekly time cards to the project manager for review and transmittal to the payroll clerk in Accounting.

The quantities reported will then be entered on the weekly labor summary run by the computer, which will show the actual production rate and average wage to date for each cost code. This is discussed in further detail in Chapter 5.

As in the processing of invoices, occasionally an invalid cost code will be reported on a weekly time card. These errors, until cleared up, will appear as Cost Code 99997, Cost Coding Errors; they are handled as discussed in Chapter 3.

CHAPTER 5

SUMMARY REPORTS

The two previous chapters have dealt with the input of the costs that comprise the work in process on the individual projects and the quantities applicable to each type of cost. With the information furnished in those sections, Accounting is then able to compile and report the total work in process of each project. The various forms to be used by the project manager comprise, as a group, the monthly job summary report. This chapter deals in detail with the various components of the monthly job summary report.

With the monthly job summary reports, the project manager may complete the project status report, to be discussed in Chapter 7. Each of the summaries report data on three cost areas:

1. *Estimated Cost.* The reports will reproduce the current estimated cost of the project in its entirety. These estimated costs are input originally through the use of the estimate listing and maintenance form, discussed in chapter 1 and Chapter 8.

2. *Actual Cost.* As mentioned above, the two reports combined represent the work in process for the project through the date shown on each of the reports. The total work in process for the project may be obtained by adding together the cumulative cost to date of the two reports. This total is reported on the project status report as discussed in Chapter 7.

Occasionally adjustments are required to actual costs, invoiced, and/or in-place quantities. These adjustments are discussed in Chapter 9, Maintenance of Actual Costs and Quantities.

3. *Project Cost.* Through the use of a computer, programmed to project the final cost of each of the summary reports, it is possible to enter the best data available during the course of construction and get the best projection of final cost. This assists the project manager in the preparation of the project status report. The procedures used in the calculation of the projected cost are discussed in detail in Chapter 6, Variance Reporting Procedures.

The job summary reports are divided into two major sections: the *sub-contract, purchase order, and materials summary* and the *labor summary*.

These two sections of the job summary reports are discussed in detail below.

SUBCONTRACT, PURCHASE ORDER, AND MATERIALS SUMMARY

The subcontract, purchase order, and materials summary constitutes the accounts payable portion of the work in process. This summary report is printed monthly in the format shown in Figure 5.1. The accounts payable is in the form of one of the three areas: subcontract, purchase order, and materials.

1. *Subcontract.* A subcontract has been defined as a contract between the general contractor and a subcontractor to provide all or a specified part of the work or materials required on a project. A sample of the sub-

```
                                         SUBCONTRACT, P.O. & MATERIAL SUMMARY FOR    0/00/00                              PAGE 00
                                                                          JOB NO & NAME  0000000    SAMPLE JOB
DIV -                          PROJ MGR- _____                         MAINTENANCE THRU FORM NO.  000
                                                        PAYABLES THRU  0/00/00

*********************************************************************************************************************************
COST CODE / COST CODE DESCRIPTION / UM  *                         *                         *                         *
----------------------------------------*             A C T U A L *           P R O J E C T E D *
          E S T I M A T E D             *                         *                         *
    QTY       UNIT COST      COST  * QTY    UNIT COST    COST   RETAIN. * QTY    UNIT COST    COST  *     VARIANCE
*********************************************************************************************************************************
  00000    SAMPLE COST CODE             *                         *                         *
  SUB.-1  0        .0000          .00   *   0     .0000      .00      .00  *  0      .0000      .00  *         .00
  MO.                                   *   0     .0000      .00      .00  *
  CUM       SUBCONTRACTOR'S NAME        *   0     .0000      .00      .00  *
  EST.BALANCE                           *                         *
                                        *
  SUB.-2                                *
  MO.                                   *
  CUM       SUBCONTRACTOR'S NAME        *
  EST.BALANCE                           *
                                        *
  P.O.-1                                *
  MO.                                   *
  CUM       VENDOR'S NAME               *
  EST.BALANCE                           *
                                        *
  MAT.                                  *
  MO.                                   *
  CUM                                   *
  EST.BALANCE                           *
                                        *
  EST.TOTAL COST               .00     * ACTUAL TOTAL COST               .00  *
*********************************************************************************************************************************


  TOTALS FOR JOB NO & NAME  0000000   SAMPLE JOB
  SUB    0                         .00    *    0                .00    *   0                .00    *        .00
  CUM                                     *    0                .00    *
                                          *
  P.O.                            .00    *    0                .00    *   0                .00    *        .00
  CUM                                     *    0                .00    *
                                          *
  MAT                             .00    *    0                .00    *   0                .00    *        .00
  CUM                                     *    0                .00    *
                                          *
  TOTAL WIP                       .00    *
  EQUIPMENT CODES                (.00)   *
  TOTAL WIP LESS EQUIP.           .00    *                                   INTEREST INCOME
```

Figure 5.1. Subcontract, purchase order, and material summary (printout).

contract agreement form is shown in the appendix. A subcontract includes essentially all labor, materials, equipment, taxes, and supervision to perform a specified portion of the prime contract between the company and the owner. In most cases, the requirements of the prime contract between the company and the owner are included as an integral part of the subcontract agreement.

There are situations in which a subcontract agreement is not issued for the performance of minor portions of the work. These would include such items as equipment rental, hauling, or miscellaneous small jobs. This work, however, still falls within the description of a subcontract item. The approval of invoices for this work has been discussed in Chapter 3. It is essential, however, that the field office secure the required certificates of insurance required by the prime contract for these subcontractors to perform work on the project.

2. *Purchase Orders.* A purchase order has been defined as an agreement, specifying items and conditions, between the company and a vendor to purchase a certain list of materials. A copy of the purchase order agreement form is shown in the appendix. A purchase order includes cost items in any code wherein *materials only* are purchased to secure firm prices on these items of material, normally including applicable sales tax and freight.

The items of material bought on a purchase order are normally included in the material column of the original estimate. Upon preparing the estimate listing discussed in Chapter 1, if the project manager has issued or intends to issue a purchase order for these materials, they should be included as purchase order items on the estimate listing. If, however, an item has been shown on the original estimate listing as material and a purchase order is subsequently awarded for it, the change in the estimate from material to purchase order is accomplished through the use of an estimate listing and maintenance form discussed in Chapter 8.

A reason for writing a purchase order for material in addition to price protection is that the computer is then able to prepare cost variances. The invoices from the vendor should be approved as purchase order items, as specified in Chapter 3.

Occcasionally, the project manager will issue a purchase order from numerous small items of material only to protect the quoted price. In these instances there is often no need for Accounting to control this document or print variances for these items. These items should still be cost coded as purchase order items with the source document input as "for file only."

3. *Materials.* The term *materials* as used in the Cost Accounting System refers specifically to those costs in accounts payable that are not covered either by a subcontract or by a purchase order. This includes items in any cost code in which materials are bought only for the project without issuing a purchase order.

The top section of the subcontract, purchase order, and material summary report provides information indicating:

1. Project manager
2. Job number
3. Job name
4. Payables through (date)
5. Division
6. Current estimated costs and quantity maintenance form number
7. Current actual estimated cost and quantity maintenance form number

Each individual cost code is listed horizontally under four major column headings with information provided vertically as follows:

1. ESTIMATED (COLUMN 1)

Line 1 provides the cost code, cost code description, and the unit of measure applicable to the cost code.

Line 2, on the left-hand side, has one of the three types of cost applicable to this section of the cost code: subcontract (SUB); purchase order (P.O.); and material (MAT).

If there are more than one of the above items in the cost code, they will be listed in the above order under the cost code. In these instances, the total of all items applicable to the cost code will be totaled on the last line of the cost code.

The estimated quantity, estimated unit cost, and estimated total cost are listed on this line horizontally.

Line 3 provides data for the monthly time period (MO) covered by the summary report and provides data to be used in the actual column of the report.

Line 4 provides cumulative (CUM) data for the project to date covered by the summary report and is used in the actual column of the report.

Also shown on the cumulative line in the estimated column is the subcontract or purchase order number applicable to this section of the cost code, along with the name of the subcontractor or vendor.

Line 5 provides the estimate balance (EST BALANCE) remaining to be spent in the estimate. Note that the estimate balance must not be confused with the actual balance to pay on a subcontract or purchase order. This balance will correspond with the balance to pay on a subcontract or purchase order only when the estimate and document amount are identical.

With the exception of the estimate balance, the information found in the estimated column of the summary report is provided by the project manager through the estimate listing, source document input, or maintenance, all of which are covered in this chapter.

SUBCONTRACT, PURCHASE ORDER, AND MATERIALS SUMMARY 27

2. ACTUAL (COLUMN 2)

Line 1 of the second column shows horizontally the source document information provided by the project manager in the form of actual total quantity, actual unit cost, and actual total cost shown on the source document. The provisions for inputting this information are described in Chapter 2. This line will be filled in only when either a subcontract or a purchase order has been issued. There will be no information provided on this line for items of material.

Line 2 of the second column shows the monthly activity of the item horizontally in the form of quantity, unit cost, and total cost for the monthly time period.

Line 3 shows the cumulative activity for the item in the form of quantity, unit cost, and actual total cost cumulative for the item.

It should be noted that the actual total cost cumulative to date is the value reported in the work in process. This amount, sometimes referred to as the billed to date amount, also includes the retainage being held on the subcontractor. The second column also shows the retainage (RETAIN) being held by the company on the source document applicable to this item.

The amount paid to date on a source document is obtained by subtracting the amount retained from the actual total cost (billed to date).

In those cost codes with multiple documents (i.e., more than one subcontract, purchase order, or material), the actual total cost (CUM) will appear at the bottom of the cost code.

The information found in the actual column of the summary report is input by the project manager through the provisions covered in Chapter 3.

3. PROJECTED (COLUMN 3)

There is only one line in the third column of the summary report. This line shows horizontally the projected final quantity, the projected unit cost, and the projected final cost of the item covered in this cost code.

The projected final quantity is input by the project manager, either in the estimate listing as explained in Chapter 1 or through the provisions discussed in Chapter 8. The projected final quantity will be shown as zero until the quantity has been proofed.

The calculations made by the computer in obtaining the projected final cost of this item of work in the cost code are explained further in Chapter 7, Variance Reporting Procedures.

4. VARIANCE (COLUMN 4)

This column reports on one line the projected cost variance of this item of work in the cost code. This cost variance is obtained by subtracting the projected final cost from the original estimated cost. It is to be noted that this is a cost variance. A plus (+) number is a cost overrun; a negative (−) is a cost underrun.

The information discussed above is provided for each item of work within each cost code in numerical sequence for the entire project.

Following the last cost code in the project are found the totals for the major items of cost for the entire subcontract, purchase order, and Material Summary.

Reported in the estimated column are the total estimated costs for:

1. Subcontract
2. Purchase order
3. Material
4. Total estimated costs
5. Total of all equipment cost codes (cost codes 01650 through 01699)
6. Total of all estimated costs less the estimated equipment costs

The totals provided in the actual column show the actual total amount of the source document, where applicable. Also shown in the actual column are the amounts paid during the monthly time period and the cumulative to date for all three items of work:

1. Subcontract
2. Purchase Order
3. Material
4. Total actual costs
5. Total of all equipment cost codes (01650 through 01699)
6. Total of all estimated costs less the estimated equipment costs

The total actual column shows the total amount of retainage being held on subcontractors and vendors on the project.

In the projected column is shown the final total projected cost for all:

1. Subcontracts
2. Purchase orders
3. Materials

The cost variance obtained by subtracting the projected total cost of each of these three categories from the total estimated cost is also shown at the bottom of the report.

The last item, on the bottom of the subcontract, purchase order, and material summary, shows the interest income earned by the project on a monthly and cumulative basis. A positive figure indicates that interest earned on excess billings received exceeds the interest expense of financing the project. A negative figure indicates that interest expense of the project is greater than the interest earned on excess billings. The sum listed represents the interest cost to the corporation to finance this deficit.

The subcontract, purchase order, and material summary provides, in a relatively concise form, the activity of each cost code on both a monthly and

SUBCONTRACT, PURCHASE ORDER, AND MATERIALS SUMMARY

a cumulative basis. Included with the subcontract, purchase order, and material summary is a supplementary report, the intent of which is to provide an itemization of the invoices processed through the monthly time period for each cost code on the individual project. This supplementary report is known as the Monthly Transaction Report. A sample of this report is shown in Figure 5.2.

The monthly transaction report is printed on the same date as the subcontract, purchase order, and material summary. The total activity on each cost code for the monthly time periods shown on the summary reports is identical to that shown for that cost code on the transaction report. Correspondingly, the total accounts payable for the monthly time period of the summary report will be identical to the total cost shown on the monthly transaction reports.

The monthly transaction report provides all of the information pertaining to each invoice within the cost code. This information is broken down horizontally on a single line as follows:

1. Subcontractor or vendor name
2. Subcontractor or vendor number
3. Invoice number
4. Project number
5. Cost Code number
6. Amount
7. Type of cost
8. Quantity
9. Unit of measure

As discussed in Chapter 3, the type of cost mentioned above is signified by one of four code references:

1. S = Subcontract
2. P = Purchase order
3. R = Retainage on either subcontracts or purchase orders
4. M = Material

VEND. #	VENDOR NAME	PO/SUB #	INVOICE #	JOB #	C.C. #	AMOUNT	INVOICED QUANTITY	U/M	SUB.	MATERIAL	P.O.
0000	SAMPLE	0000000	000			.00			.00	.00	.00
		COST CONTROL TOTAL									
		COST CONTROL TOTAL									
	SAMPLE JOB					.00	TOTAL THIS JOB		.00	.00	.00

MONTHLY TRANSACTION REPORT   DATE RUN 0/00/00   PAGE 00

Figure 5.2. Monthly transaction report (printout).

When an invoice has been processed for material the total amount of the invoice will be shown and will be followed immediately to the right by the letter M.

When an invoice is processed on a subcontract or purchase order on which retainage is being held, the processing of that invoice will be shown on two separate lines of the transaction report. The first line will indicate the retained amount of the subcontractor's or vendor's invoice. This will be followed immediately to the right by the letter R. Immediately below the line for retainage will be a second line for the same invoice that will indicate the amount paid to the subcontractor. Immediately to the right of this will be either the letter S or the letter P. The total of the retainage plus the amount paid on the invoice will be the amount billed by the subcontractor. This amount is the total work in process chargeable by the invoice.

If an invoice is processed on a purchase order or subcontract with no retainage being held, the invoice will be shown only on one line with the total amount of the invoice being shown followed by P or S. Immediately to the right of this will be the total quantity for that monthly time period. To the right of this is the total breakdown of cost within the cost codes for either subcontract, purchase order, or material.

LABOR SUMMARY

Unlike the subcontract, purchase order, and material summary, which is printed only on a monthly basis, the labor summary is printed both weekly and monthly. The reason is that accurately monitoring the production rate within a cost code often requires a weekly review of the activity.

The weekly labor summary will print only those cost codes with activity for that week. The weekly labor summary includes only estimated costs and actual costs, and does not include calculations for projected costs or cost variances. (See Figure 5.3.)

The monthly labor summary will print all cost codes estimated for the project, in order that a complete project status report may be prepared. (See Figure 5.4.)

As in the subcontract, purchase order, and material summary, the top portion of both the weekly and monthly labor summary reports provides the following information:

1. Project number
2. Project name
3. Project manager name
4. Division
5. Payroll through date
6. In-place quantity through date
7. Maintenance form number included

LABOR SUMMARY   31

```
                    WEEKLY LABOR SUMMARY                                    WEEK ENDING 0/00/00              PAGE 00
         DIV.                              PROJ.MGR.                JOB # & NAME  000000
    COST CODE      COST CODE DESCRIPTION    UM    QUANTITY  AVG.WAGE  PROD. U/MH  PROD. MH/U   TOT MAN HRS   TOTAL COST
    00                                      LS      .00     .000       .0000       .0000         .00          .00
    00                                      LS      .00     .000       .0000       .0000         .00          .00
                   JOB TOTALS                       .00                                          .00          .00
```

Figure 5.3. Weekly labor summary.

The labor summary reports are prepared through the input provided by the project manager and superintendent through the weekly time cards and unit completion reports. These have been discussed in Chapter 4.

The arrangement of the monthly labor summary is slightly different from that of the subcontract, purchase order, and material summary. The report is divided into only three vertical columns:

Column 1 of the monthly labor summary is composed of both the estimated and actual data.

Column 2 is the projected cost data.

Column 3 is the projected cost variance data.

The weekly labor summary is run in numerical cost code sequence with the top line in each cost code being the cost code number, cost code description, and the unit of measure of the cost code. This information runs horizontally.

Line 1 within the cost code is the estimate (EST). The information included across this line is the estimated quantity, estimated average wage, estimated production rate, estimated total labor hours, and, finally, the estimated total cost. This information is provided through the input of the project manager as discussed in Chapter 1.

Line 2 within the cost code is the weekly (WKLY) activity. This will show across the line the activity within the cost code for the week through the payroll date. This will include the weekly quantity, average wage, production, labor hours, and total cost.

Line 3 within the cost code is the cumulative (CUM) activity. This is the total activity for the cost code cumulative through the date of the report.

```
                                              MONTHLY LABOR SUMMARY                                           PAGE 00
DIV -                              PROJ MGR-              JOB NO. & NAME 0000000  SAMPLE JOB
                                        IN PLACE QTY THRU  0/00/00     PAYROLL THRU  0/00/00    MAINTENANCE THRU FORM NO.  00
*********************************************************************************************************************
CODE CODE / COST CODE DESCRIPTION     /UM              *                                   *
---------------------------------------------------    *   PROJ.   PROJ.   PROJECTED        *   PROJECTED
                                                       *   FINAL   AVG.   PROD.   PROD. TOTAL  FINAL  *   PROJECTED
    QTY.   AVG.   PROD.   PROD.  * TOTAL    TOTAL     *   QTY    WAGE   U/MH    MH/U   MAN HRS   COST   *   VARIANCE
           WAGE   U/MH    MH/U     MAN HRS   COST      *                                    *
*********************************************************************************************************************
  00000  SAMPLE COST CODE              LS              *                                   *
  EST.    0   .000   .0000   .0000   0     .00        *                                   *
  WKLY    0   .000   .0000   .0000   0     .00        *                                   *
  CUM     0   .000   .0000   .0000   0     .00        *   0    .000   .0000   .0000   0    .00   *    .00
  EST.BALANCE                                          *                                   *
*********************************************************************************************************************

   J O B   T O T A L S   QTY.      MAN HRS       COST      BALANCE   PROJ FINAL QTY   PROJ MAN HRS   PROJ COST   COST VARIANCE
               EST.       .00        .00         .00         .00
               WKLY       .00        .00         .00
               CUM        .00        .00         .00                     .00              .00          .00           .00
```

Figure 5.4. Monthly labor summary (printout).

Included will be the cumulative quantity, average wage, production rate, labor hours, and cost.

Line 4 within the cost code is the estimate balance (EST BALANCE). As in the subcontract, purchase order, and material summary, this estimate balance represents the amount remaining from the estimated cost to be spent in the cost code.

The projected section of the summary report provides on a single line the following calculations: projected final quantity, projected average wage, projected production rate, projected total labor hours, and projected final cost. The calculations that are made to obtain the projected final cost are discussed in further detail in Chapter 6.

Line 5, the final vertical column of the monthly labor summary report, is the projected cost variance. This is obtained by subtracting the projected final cost from the estimated cost. As discussed earlier, a plus (+) number indicates a cost overrun and a negative (−) number indicates a cost underrun.

One item of labor cost not normally associated with production rates and average wage is cost code 01802, Labor Tax. As labor tax is directly related to the cost of labor, it is included within the labor summary report.

Line 1 of this cost code is the total estimated labor tax on the project.

Line 2 represents the labor tax applicable to the labor for that weekly period.

Line 3 represents the cumulative labor tax applicable to the project.

As in the other cost codes an estimate balance is shown.

As the labor tax applicable to a project is directly related to all other labor costs, no projected final cost calculations are shown in Column 2. As all other projected final costs are provided as a basis for the project manager to prepare the project status report, it is assumed that many will change, some significantly. It is, therefore, unreasonable to assume that a projected final cost for labor tax could be forecast that would be accurate.

CHAPTER 6

VARIANCE REPORTING PROCEDURES

As discussed in Chapter 5, the subcontract, purchase order, and material summary and the labor summary have distinct characteristics that warrant the two reports. For the same reason, the cost variance reporting procedures for each of the two reports are also distinctively different.

A computer program has been prepared to make cost variance calculations to assist the project manager in the preparation of his or her project status report, to be discussed in Chapter 7. Similar calculations would be made by the project manager in computing the actual variance anticipated with the personal input that is obviously required in each cost code.

The variance as shown on the computer and as calculated by the project manager is based on *cost*. A plus (+) is a cost overrun, and a negative (−) is a cost underrun.

If the projected final quantity has not been input by the project manager, the computer will forecast from the original estimated quantity. If the projected final quantity is the same as the original estimated quantity, then this information should be input through the use of an estimate listing and maintenance form, in order to show that this quantity has been proofed. Otherwise, projected final quantity will show zero (0) on the summary reports.

If a cost code has no activity and the quantity *has not been proofed*, then the projected final cost will be shown as the original estimated cost.

If, however, a cost code has no activity, and the quantity *has been proofed* and is different from that of the original estimated quantity, then the projected final cost will be calculated on the projected final quantity and the original estimated production rate and average wage.

If a cost code has had activity and there are missing data such as estimated quantity, then the report will show zero (0.00) projected final cost and will print "Missing Data" in the cost variance column. This is to alert the project manager to a potential error in the report.

In some instances the project manager may find that he or she is required to use a cost code with no estimated costs. In this situation the project manager should report on the estimate listing and maintenance form only the projected final quantity. If an original estimated quantity is shown without an original estimated cost, it will result in a "Missing Data" in the cost variance column. When this occurs, updating of the projected final quantity through the use of an estimate listing and maintenance form will be required.

Once again, it must be remembered that this is a cost variance. A plus (+) variance is a cost overrun; a negative (−) is a cost underrun.

The calculations made by the computer in obtaining the cost variances for the two summary reports are explained separately below.

SUBCONTRACT, PURCHASE ORDER, AND MATERIAL SUMMARY

The input of invoiced quantities and the approval of all invoices have been detailed in Chapter 3 and should be reviewed in considering the calculations of the cost variances of the subcontract, purchase order, and material summary.

LUMP SUM SUBCONTRACTS AND PURCHASE ORDERS

Those items for which the project manager has issued a lump sum source document show the projected final cost as the actual total cost shown on the source document, including change orders. The projected final cost minus the estimated cost will be cost variance.

Since quantity does not pertain to lump sum subcontracts and purchase orders, the input of quantity is not necessary.

When a lump sum subcontract or purchase order is submitted to Accounting, "Lump Sum" should be noted in the upper right-hand corner along with the additional information covered in Chapter 2.

UNIT PRICE SUBCONTRACTS AND PURCHASE ORDERS

For those subcontracts and purchase orders that are written only for a *unit price*, the following calculations will be made:

1. The projected final quantity multiplied by the source document unit price being paid will give the projected final cost.
2. The projected final cost minus the estimated cost will be the cost variance.
3. The unit price of the contract will be shown in the actual column of the report on Line 1 of the subcontract, purchased order, and material summary.

There are several cost codes in the subcontract, purchase order, and material summary in which the unit of measure is lump sum, such as

Office and Storage Trailers, Water, Power, and so on. In order to allow the computer to forecast a variance in the items, the quantity of 100 (representing 100%) is to be input on the estimate listing. Then with each invoice approved, input the percent of completion that each invoice represents. As an example:

1. Cost Code 01603, Power. The estimate for the 10-month project is a lump sum of $750.
2. On the estimate listing $750 is input as the estimated cost and 100 (100%) as the quantity.
3. The first monthly invoice is for $50. Since it is early in the job, it is estimated that this invoice only represents 5% of the total. Therefore, input a quantity of 5 on the invoice approval form.
4. The second monthly invoice is for $75, and it is estimated that this represents 10% of the amount of power to be used. Therefore, a quantity of 10 is input on the invoice approval form.

If during construction, the project manager realizes a cost variance is to occur in a cost code, the projected final quantity may be updated by changing the quantity (representing percentage) of the cost variance. As an example:

1. The above estimate of $750 on Cost Code 01603, Power, did not include a $150 temporary service charge to the power company.
2. The projected final quantity can be increased by 20 (representing 20%). The revised quantity is therefore 120 (representing 120%).
3. The projected final cost will then be forecast by the computer as $900, representing a cost variance of $150.

LABOR SUMMARY

The input of in-place quantities and the cost of labor have been detailed in Chapter 4. These items should be reviewed in considering the calculations for cost variances in the labor summary.

To obtain the projected cost variance on the labor summary report, the computer has been programmed to make the following calculations:

1. The projected final quantity divided by the cumulative productions U/MH (or multiplied by MH/U) will give the projected total labor hours.
2. The projected total labor hours multiplied by the cumulative actual average wage will give the projected final cost.
3. The projected final cost minus estimated total cost will give the projected cost variance.

As in the subcontract, purchase order, and material summary, there are several cost codes in the labor summary in which the unit of measure is lump sum, such as Office and Field Supervision or Field Engineering. In order to allow the computer to forecast a variance in the items, input the quantity of 100 (representing 100%) on the estimate listing. Then input the percent complete of that item with the unit completion report. The computer will then be able to project a cost variance in these cost codes.

Please keep in mind that the computer cost variances are intended only as an aid in the preparation of the project status report and each cost code must be reviewed for accuracy.

CHAPTER 7

PROJECT STATUS REPORT

Each month a general contractor should prepare a financial report for the entire corporation. To facilitate compilation of this report it is essential that the status of each project be accurately reported.

To accomplish this, the project manager is to complete monthly the *project status report*. A copy of this report is shown in Figure 7.1.

The importance of the project status report cannot be overemphasized. The need for accurate information in this report is, to a large degree, the reason for the Cost Accounting System.

The monthly preparation of the project status report affords the project manager the opportunity to analyze several of the essential ingredients that determine whether or not a project has been successful for the corporation. These include the completion date, the anticipated profit, the billing structure, and the owner's relationship with the general contractor. It is reasonable to assume that if a project is on schedule, within the budget, and built to the owner's satisfaction, quality will also be satisfactory.

An explanation of the various sections of the report are detailed below.

PROJECT STRUCTURE

Lines 1 through 5 in this section of the report analyze the contract status with reference to contract amount, anticipated cost, anticipated profit, and the percent of profit anticipated on the project.

Line 1, original contract, represents the original estimate. The anticipated cost on Line 1 is to be identical to the total of the Estimate Listing. The total contract amount minus the anticipated cost gives the anticipated profit of the original contract. The original anticipated profit divided by the original anticipated cost results in the original percent of profit on the project.

PROJECT STATUS REPORT #_____   DATE:_____
                                   JOB NO:_____

PROJECT:_____
PROJECT MANAGER:_____ SUPERINTENDENT:_____

PROJECT STRUCTURE	TOTALS	ANTICIPATED COST	ANTICIPATED PROFIT	PROFIT COST %
1. Original Contract				
2. Change Orders (Thru C.O.#__)				
3. Sub-Total				
4. P & L Forecast				
5. Total				

BILLING STRUCTURE

6. Total Contract Billed _____   % Complete: Line $\frac{8}{5}$ = _____ %
7. Less Contract Retainage _____
8. Less Work In Process _____   % Complete: Line $\frac{12}{5}$ = _____ %
9. Plus W.I.P. Retainage _____   (Cost)
10. Total Overdraw _____   % Overdraw: Line $\frac{10}{8}$ = _____ %
11. Volume of Subs Bonded _____
12. W.I.P. (Less stored mat'ls.)_____   % Time Used: Line $\frac{19}{21}$ = _____ %
13. Interest This Month _____
14. Interest Cumulative _____

BUDGET ANALYSIS	ANTICIPATED COSTS	W.I.P.	%
15. Labor			
16. SC - PO - MAT(Less equip.)			
17. Equipment			

COMPLETION STATUS	DATE	DAYS	OWNER'S REPRESENTATIVE STATUS	
18. Original Contract			Excellent Relationship	☐
19. Elapsed Time			Cooperates	☐
20. Original Completion			Questionable Attitude	☐
21. Extended Completion			Difficult Relationship	☐
22. Estimated Completion				

PROJECT MANAGER'S REMARKS   W.I.P.-Mo.$_____ Requested C/O$_____ Days_____

_____
_____

DIVISION MANAGER REMARKS _____
_____

Due Second Working Day After Receipt of Monthly Job Summary

Figure 7.1. Project status report.

Since Line 1 represents the original estimate, it should never be changed.

Line 2, change orders, represents the change orders *executed* by the owner. The anticipated cost on Line 2 is the cumulative costs that are input into the system through the use of the estimate listing and maintenance form, explained in Chapter 8. The cost of the change orders on Line 2 must be kept current through the use of the maintenance form. If a lag exists in either this report or the summary reports, incorrect forecasting of the individual accounts will inevitably occur. As in Line 1 of the report, the total amount of change orders minus the anticipated cost represents the cumulative percent of markup obtained through the change orders.

Line 3, the subtotal of Line 1 and Line 2, represents the total estimated line. It is from this line that the profit and loss (P & L) forecast is to be made. The anticipated cost on this line *must* be kept current with the total estimated costs shown on the monthly summary reports. This is done through the use of the estimate listing and maintenance form as explained in Chapter 8. This item should be checked each month prior to performing the P & L forecast on the individual accounts. Discrepancies should be noted prior to preparation of the P & L forecast and corrections made prior to the next month.

Line 4, P & L forecast, is the *most important line* on the report. It is also considered one of the most important duties of the project manager.

This line represents the projected cost variance from the estimated cost. The two columns to be filled in on Line 4 should always remain in balance. An additional (+) cost results in an equal decreased (−) profit, and vice versa.

Each cost code as summarized on the summary reports is to be analyzed as to the true anticipated cost. It is not expected that this report be either optimistic or pessimistic. This should be a realistic forecast of the anticipated P & L on the project.

In addition, the project manager is to analyze each valid area of risk management. Any work item that contains a distinct possibility of incurring a cost overrun that will materially affect the P & L forecast is to be considered an anticipated cost expenditure until such time as the risk no longer exists. As an example, a deep excavation is still required on the project, and dewatering may be necessary. If that is the case, the estimated cost for this item of $5,000 is to be included as a projected cost until after the excavating is complete. These items falling into the area of risk management are to be listed in the remarks portion of the report.

It is assumed that the P & L forecast for each account will be prepared directly on the monthly summary reports. This can be accomplished simply by marking through the projected final cost and the projected cost variance as calculated by the computer, and substituting the project manager's estimates for those columns.

Line 5, total, represents the total contract status. It includes the total of the original contract, the change orders, and the P & L forecast. Once again, the total contract amount minus the anticipated cost represents the anticipated profit on the project. Anticipated profit divided by anticipated cost represents the final percent of profit anticipated on the project.

BILLING STRUCTURE

Lines 6 through 14 of this section analyze the cost of the project to date with reference to the receivables or billings. The billing structure of all projects under construction, when analyzed with the cost of operating the corporation, reflects the cash flow status of the entire company. Needless to say, this is an essential consideration in the preparation of each progress billing.

Line 6, total contract billed, represents the gross contract billing to date.

Line 7, less contract retainage, represents the amount of retainage withheld by contract by the owner on each progress billing.

Line 8, less work in process (WIP), represents gross cost to date on the project. This line may be filled in by adding the totals of the two monthly summary reports.

Line 9, plus WIP retainage, represents the amount of retainage held by the general contractor on the subcontractors for the project. This line may be filled in from the total column of the subcontract, purchase order, and material summary for retainage.

Line 10, total overdraw, represents the net total of Line 6 through Line 10. This net figure represents the total billings in excess of cost (overdraw), or vice versa should the project be in a negative cash position.

Line 11, volume of subs bonded, represents current listing of any subcontractors that have furnished performance and payment bonds on the project, including any change orders to date.

Line 12, WIP (less stored materials), represents Line 8 (work in process) less the estimated amount of stored materials that are currently included in the work in process total.

Line 13, interest this month, represents the interest earned on the project through overdraw for the previous month. A negative number indicates an adverse cash flow requiring corporate financing of the project.

Line 14, interest cumulative, represents the cumulative interest earned, through overdraw, for the project. The interest this month and the interest cumulative are shown each month at the bottom of the subcontract, purchase order, and material summary.

The right side of the billing structure section provides several key percentages, including:

Percent Complete. Line 8 (work in process) divided by Line 5 (anticipated cost) represents the percent complete of the project as analyzed by the accounting system.

Percent Complete. Line 12 (work in process less stored materials) divided by Line 5 (anticipated cost) represents the actual in-place percent complete of the project.

Percent Overdraw. Line 10 (total overdraw) divided by Line 8 (work in process) represents the percent of overdraw of the project.

Percent Time Used. Line 19 (elapsed time—days) divided by Line 21 (extended completion days) represents the percent of time used on the project.

BUDGET ANALYSIS

Lines 15 through 17 of this section serve as a comparison of the final anticipated cost of the project to the current work in process. The three categories of anticipated costs are to total the final anticipated cost projected on Line 5 of the report. The three categories of work in process are to total Line 8 of the report. The percent of work in process divided by anticipated cost in each category reflects the percent of cost expended to date.

Line 15, labor, represents anticipated cost for labor, a projection obtained by the project manager in the analysis of the monthly labor summary. The work in process amount may be obtained directly from the totals shown at the bottom of the monthly labor summary.

Line 16, subcontractor—purchase orders and materials (less equipment), represents the anticipated cost for this area, a projection obtained by the project manager in the analysis of the monthly subcontract, purchase order, and material summary, less the anticipated cost of equipment to be used on the project. The work in process amount may be obtained by subtracting the equipment cost to date from the totals shown at the bottom of the monthly subcontract, purchase order, and material summary.

Line 17, equipment, represents the anticipated cost of equipment that is the project manager's projection of cost codes 01650 through 01699 of the subcontract, purchase order, and material summary. The work in process amount is the total expended to date in these same cost codes.

COMPLETION STATUS

Lines 18 through 22 of the report are designed to give a summary of various time factors involved on the project, in terms of actual calendar dates and also the number of calendar days attached to each category.

Line 18, original contract, represents the date, usually the effective date of the notice to proceed, taken from the original contract. The number of calendar days allowed under the original contract is also to be noted.

Line 19, elapsed time, represents the number of calendar days that have elapsed since the notice to proceed.

Line 20, original completion date, represents the original contract completion date, along with the number of calendar days allowed by the original contract.

Line 21, extended completion date, represents the current completion date required by the contract, along with the total number of calendar days allowed to complete the contract under the extended time. This is to be *only* the extensions that have been approved in writing. Pending extensions are not to be considered.

Line 22, estimated completion, represents the date that the project manager is currently estimating for completion of the project. This is *not* the contract completion date. This date is the *estimated* completion date. In addition, the total number of days estimated to complete the project from the construction start date is to be noted.

OWNER'S REPRESENTATIVE STATUS

The purpose of this section of the report is to give consideration to the relationship between the general contractor and the owner. This area often is directly related to the current estimated completion date, the profit, the quality of the project, or a combination of these items. A brief explanation of any unusual relationship with the owner is anticipated.

PROJECT MANAGER REMARKS

In addition to any notes the project manager needs to make regarding the report in general, there are three items that are to be completed each month:

1. Work in process that has been run off since the last month's report. This can be obtained by subtracting Line 8 of the current report from the previous month's report.
2. The amount of any requested change orders that are pending with the owner. This is not approved change orders, but only those that have been submitted but not yet approved.
3. The number of calendar days of time extension that have been requested, but that are still pending approval by the owner.

DIVISION MANAGER REMARKS

This section is reserved for use by division managers.

CHAPTER 8

MAINTENANCE OF ESTIMATED QUANTITIES AND COSTS

It is essential that a Cost Accounting System developed for use in construction provide a method for input of the various *changes* that occur during construction of the project. In order to maintain the status of each project on a current basis, the estimate listing and maintenance form enables input to be submitted to Accounting. This form serves a dual purpose, first as an estimate listing, as discussed in Chapter 1 and second as a maintenance form by crossing off *Estimate Listing* in the title.

Examples for uses of the maintenance form are found in Figure 8.3 through Figure 8.6. The estimate listing and maintenance forms are numbered in sequence for each project, beginning with the estimate listing.

The project manager submits the completed maintenance form to the accounting supervisor. This form is processed through Accounting and Data Processing before it is returned to the project manager.

Accounting returns the original maintenance form to the project manager, along with a data run updating the cost codes that have been changed. This data run is called the Maintenance Update. A copy of this maintenance update is shown in Figure 8.1.

The maintenance form, shown in Figure 8.2, reflects only the net amount of changes. That is, when a change is made in the status of any item in a cost code, the project manager fills in only the net increases (+) and decreases (−) in the amounts added or deducted from the current data shown on the summary reports.

The three major uses for the maintenance form are: input of projected final quantities, change orders, and net redistribution of estimated costs.

INPUT OF PROJECTED FINAL QUANTITIES

This represents input of the projected final quantity on each item estimated other than a lump sum item. Only through the input of accurate

MAINTENANCE OF ESTIMATED QUANTITIES AND COSTS

```
                                                                    DATE RUN 0/00/00              PAGE    00
        MAINTENANCE UPDATES - ESTIMATES                                              SYSTEM MAINTENANCE FORM NO.
DIV. _____    PROJ.MGR. _____    JOB NO. & NAME 0000000    SAMPLE JOB
*********************************************************************************************************
              SYSTEM MAINTENANCE DATA ENTERED THIS FORM       *    CUMULATIVE ESTIMATE DATE INCLUDING THIS FORM
                                                              *                                              EST UNIT COST
COST CODE     DOC TYPE-#   UM  *                              *                                                   OR
------------------------------ *              EST.            *                            EST.                AVG WAGE
COST CODE DESCRIPTION         * EST. QTY.  EST.MAN HRS  TOT COST   PROJ. QTY  * EST.QTY.  EST.MAN HRS  TOT COST  PROJ QTY
00000                                                         *
SAMPLE COST CODE DESCRIPTION    .00         .00         .00         .00       *  .00         .00         .00        .00

    NET MAINT.FORM ADJUST       .00         .00         .00         .00          ERRORS THIS FORM              00000
```

Figure 8.1. Maintenance update—estimates (printout).

proofed quantities can reasonable variance reporting be obtained. Input is submitted on the maintenance form as follows:

1. Fill in job name and number.
2. Fill in maintenance form number.
3. Fill in cost code number.
4. Fill in document type and number, such as S0851 (for subcontractors), or SUB (where no subcontract is to be written), or P2811 (for purchase orders), or material, or labor.
5. Fill in projected final quantity.
6. Total the estimated labor hours and cost columns.

CHANGE ORDERS

This item is to be used whenever a change order is issued by the owner and is to be input upon receipt of the executed change order.

This item is to be used to make the net change required in an item and not used simply to adjust the originally estimated item. Also, reflected in this would be the change to the project quantity. Use the maintenance form as follows:

1. Fill in project name and number.
2. Fill in maintenance form number.
3. Fill in change order number.
4. Fill in cost code number.
5. Fill in document type and number.
6. Show net increase (+) or decrease (−) to estimated quantity.
7. Reflect only change in unit of measure.
8. Show net increase (+) or decrease (−) to labor hours.
9. Show net change to total estimated cost.
10. Show net change to projected final quantity.
11. Total the estimated total labor hours and cost columns.

CHANGE ORDERS 45

ESTIMATE LISTING & MAINTENANCE FORM

JOB NAME_____ JOB NO._____ MAINTENANCE FORM NO._____

COST CODE	DOCUMENT TYPE #	ESTIMATED QUANTITY	U/M	ESTIMATED TOTAL LH	ESTIMATED TOTAL COST	PROJECTED QUANTITY

TOTALS
Submitted by:_____ Date:_____
Processed by (Accounting):_____ Date:_____   PAGE____OF____
Processed by (Data Process.):_____ Date:_____

Figure 8.2. Estimate listing & maintenance form (blank form).

As an example, assume the owner has requested that the general contractor (1) modify the site as per owner's plan, and (2) delete 115 LF of track. The estimate for this work is:

1. Modify the site plan:
   a. Cost Code 02210, Site Grading, the subcontract cost
      to add 4300 cubic yards of fill   = $5,000.00
   b. Cost Code 01713, Field Engineering,
      Labor to lay out revised site plan:
      60 labor hours × $7.00/hr.   =   420.00

c. Cost Code 01802, Labor tax,
$420.00 \times 11\%$ (or as applicable)=$46.20   47.00

Subtotal cost Item 1	$5,467.00
Overhead and profit at 62.2% =	$3,400.00
Total change order	$8,867.00

2. Delete 115 LF of track:
   a. Cost Code 02610, Paving, deduct subcontract amount for 306 square yards of Type I paving: (−)   (900.00)
      Overhead and profit   72.00
      Total change order Item 2
        Deduct   (828.00)
      Subtotal change order =
        Add $8,039.00
3. Add:
   Cost Code 01805, General Contract Bond,
   $8,150.00 \times 0.01$ (or as applicable) = $81.50
                                                              82.00
   Total change order $8,121.00

When the signed change order is received from the owner, the project manager is then responsible for inputting the costs on a maintenance form. The following example demonstrates the appropriate items to be completed for inputting these costs. Figure 8.3 illustrates maintenance form number one. The entries required on the maintenance form number two would be as detailed below.

1. Project name—Sample High School
2. Project number—4068
3. Maintenance form number—1
4. Change order number—1
5. Cost Code 02210, Site Grading
   a. Document type and number = S0820
   b. Estimated quantity = +4,300 CY
   c. Estimated total cost = +$5,000.00
   d. Projected final quantity = +4,300 CY
6. Cost Code 02610, Paving
   a. Document type and number = S0820
   b. Estimated quantity = (306 SY)
   c. Estimated total cost = ($900.00)
   d. Projected final quantity = (306 SY)
7. Cost Code 01713, Field Engineering
   a. Estimated quantity = N/C

CHANGE ORDERS

ESTIMATE LISTING & MAINTENANCE FORM

JOB NAME SAMPLE HIGH SCHOOL   JOB NO. 4068   MAINTENANCE FORM NO. 1
                                                      CHANGE ORDER NO.

COST CODE	DOCUMENT TYPE #	ESTIMATED QUANTITY	U/M	ESTIMATED TOTAL LH	ESTIMATED TOTAL COST	PROJECTED QUANTITY
02210	S0820	4300			5,000	4300
02610	S0820	(306)			(900)	(306)
01713	LABOR			60	420	
01802	LABOR				47	
01805	MAT				82	
TOTALS				60	4,649	

Submitted by:_____XXXX_____ DATE:___XXXX___
Processed by (Accounting):_____ DATE:_____   PAGE 1 OF 1
Processed by (Data Process.):_____ DATE:_____

Figure 8.3. Maintenance form 1.

    b. Estimated labor hours = + 60 LH
    c. Estimated total cost = +$420.00
    d. Projected final quantity = N/C
8. Cost Code 01801, Labor Tax
    a. Estimated total cost = +$47.00
9. Cost Code 01805, General Contract Bond
    a. Estimated total cost = +$82.00

The input information above deals only with the *cost* of the change. The profit is reflected only in the difference between cost and the receivables.

MAINTENANCE OF ESTIMATED QUANTITIES AND COSTS

Occasionally, a situation arises wherein work must be performed on a project that is either (1) *not* reimbursable from the owner, or (2) omitted from the original estimate by error. In these cases, *no* change is to be made on the maintenance form.

However, change orders may have to be issued to subcontractors and vendors, and additional costs may be incurred in materials and labor by the general contractor. When the source document, material, and labor costs are processed through Accounting, the actual total cost subtracted from the estimated cost will reflect the cost variance.

The only change allowed would be to update the projected final quantity on items other than lump sum.

NET REDISTRIBUTION OF ESTIMATED COSTS

This item refers to transfer of estimated cost from one category to another, such as:

1. Subcontract to purchase order, material, and/or labor
2. Subcontract to labor only
3. Material to purchase order
4. Termination of a subcontract and issuing of a new subcontract to a different vendor

Changes in 1 and 2 could occur either before the subcontract was issued or through termination after the subcontractor has been partially paid. Note that these changes should always net to zero. Also, the reverse of any of these changes could occur.

An example shown in maintenance form number two, in Figure 8.4, involves a decision to perform the painting on a project with the general contractor's own material and labor, rather than subcontracting as originally bid and budgeted; the decision was made before a subcontract was issued. The budgeted costs, therefore, must be deleted from the subcontract line and redistributed to purchase order, labor, and labor tax.

```
Original estimate:
    Delete Sub,
                            100,000 SF @              $20,000
New estimate:
    Add purchase order
                            100,000 SF @              $3,000.00
    Add labor
    100,000 SF at 2,982 LH                            14,167.00
    Add labor tax on                                   2,833.00
    ($15,315 @ 20%) (as applicable)
    Total cost:                                       $20,000.00
    (same as subcontract amount deleted)
```

NET REDISTRIBUTION OF ESTIMATED COSTS

ESTIMATE LISTING & MAINTENANCE FORM

JOB NAME SAMPLE HIGH SCHOOL   JOB NO. 4068   MAINTENANCE FORM NO. 2
                                                  CHANGE ORDER NO. _____

COST CODE	DOCUMENT TYPE #	ESTIMATED QUANTITY	U/M	ESTIMATED TOTAL LH	ESTIMATED TOTAL COST	PROJECTED QUANTITY
09900	SUB-1	-100,000			-20,000	
09900	P4446	+100,000			3,000	
09900	LABOR	+100,000		+2,982	14,167	
01802	LABOR				2,833	
TOTALS				2,982	-0-	

Submitted by: ____XXXX_____ DATE: __XXXX____
Processed by (Accounting): _____ DATE: _____   PAGE 1 OF 1
Processed by (Data Process.): _____ DATE: _____

Figure 8.4. Maintenance form 2.

Maintenance Form number three, in Figure 8.5, involves the same item, except in this case subcontract 0326 has been issued and 30,000 SF at $6,000 invoiced by the subcontractor before the decision was made to cancel the subcontract and complete the work with the general contractor's own work force. This requires that the uncompleted portion of the subcontract be deleted and then redistributed to purchase order, labor, and labor tax.

Uncompleted original estimate
    Delete Sub,
                          70,000 SF (100,000-30,000) @ $14,000.00

New Estimate
    Add purchase order
        70,000 SF @   2,100.00
    Add labor
        70,000 SF @ 2,088 LH   9,916.00
    Add labor tax on
        $10,439 at 20% (as applicable)   1,984.00
    Total cost $14,000.00
    (same as subcontract amount deleted)

NOTE: 30,000 SF at $6,000 will remain on subline.

ESTIMATE LISTING & MAINTENANCE FORM

JOB NAME   SAMPLE HIGH SCHOOL   JOB NO.   4068   MAINTENANCE FORM NO.   3
                                                                             CHANGE ORDER NO.

COST CODE	DOCUMENT TYPE #	ESTIMATED QUANTITY	U/M	ESTIMATED TOTAL LH	ESTIMATED TOTAL COST	PROJECTED QUANTITY
09900	S0326	-70,000			-14,000	
09900	P0521	+70,000			+ 2,100	
09900	LABOR	+70,000		2,088	+ 9,916	
01802	LABOR				+ 1,984	
TOTALS				2,088	-0-	

Submitted by: XXXX   DATE: XXXX
Processed by (Accounting):   DATE: PAGE   OF
Processed by (Data Process.):   DATE:

Figure 8.5. Maintenance form 3.

NET REDISTRIBUTION OF ESTIMATED COSTS

ESTIMATE LISTING & MAINTENANCE FORM

JOB NAME __SAMPLE HIGH SCHOOL__   JOB NO. __4068__   MAINTENANCE FORM NO. __4__

CHANGE ORDER NO. _____

COST CODE	DOCUMENT TYPE #	ESTIMATED QUANTITY	U/M	ESTIMATED TOTAL LH	ESTIMATED TOTAL COST	PROJECTED QUANTITY
03300	MAT		LS		-30,000	
03300	P1680		LS		+30,000	
TOTALS					-0-	

Submitted by: __XXXX__   DATE: __XXXX__
Processed by (Accounting): _____   DATE: _____   PAGE __1__ OF __1__
Processed by (Data Process.): _____   DATE: _____

Figure 8.6. Maintenance form 4.

Termination of a subcontract for labor only with the labor to be performed by the contractor would be the same as above without any transfer to purchase order.

Termination of a subcontract and issuance of a new subcontract would follow the same principle outlined in examples from maintenance forms two and three. This could occur before any work started on the initial subcontract or after partial performance.

In addition to the maintenance form, Accounting is given cancellation of the original subcontract along with a copy of the purchase order for the material.

MAINTENANCE OF ESTIMATED QUANTITIES AND COSTS

In many instances, costs may be budgeted under the material line on the original entry and a later decision made to issue a purchase order for all or part of the item.

An example would be from material estimated at $100,000 total budgeted under Cost Code 03300 as material. In the original estimate, $30,000 of this amount was for plywood. After negotiations were complete, purchase order 1680 was issued for $20,000 covering all the plywood. It will now be necessary to transfer the estimated plywood cost from material to purchase order leaving $70,000 under material. Maintenance form number four, shown in Figure 8.6, demonstrates how this would be handled.

The entry would be as follows:

Original estimate
        Delete material   LS $30,000.00

New estimate
        Add purchase order   LS $30,000.00

In addition to the maintenance form, Accounting is given a copy of the purchase order.

Note that when Accounting processes the purchase order, the system will show a favorable variance of $10,000 (difference between $30,000 estimate and $20,000 actual cost).

CHAPTER 9

MAINTENANCE OF ACTUAL COSTS AND QUANTITIES

Inevitably, at some time during the construction of a project, it will be necessary to update the cost of the project. Adjustments are made through the use of the Actual Cost and Quantities Maintenance Form, described below and shown in Figure 9.1.

Acutal cost adjustments can occur for:

1. In-place quantity
2. Labor cost
3. Invoiced quantity
4. Subcontract, purchase order, and material cost

Adjustments to in-place and invoiced quantities *only* may be submitted to Accounting directly by the project manager.

Adjustments to labor cost and subcontract, purchase order and material cost require division manager approval. Adjustments to cost within a project obviously must balance out to zero. That is, it is possible to transfer cost from one cost code to another but the total project cost must remain the same.

No cost adjustments transfer between projects may occur without the president's approval.

The use of the actual cost and quantities maintenance form is shown in Figure 9.1 and explained in four examples below.

Example 1. A material invoice is coded to a cost code that is invalid for the project. An invoice is approved by the project manager for 10 CY of concrete at a total cost of $312. Instead of being charged to Cost Code 03020, Foundation Concrete, the invoice was approved to Cost Code 03002, which does not exist. Rather than stop the entire accounting process, the invoice is charged by Accounting to Cost Code 9999, Cost Code Error. This allows the accounting process to continue and shows the error immediately to the project manager.

ACTUAL COST AND QUANTITY MAINTENANCE FORM

JOB NAME ___SAMPLE JOB___   JOB NUMBER ___XXXX___   FORM NUMBER ___XX___

COST CODE	DOCUMENT TYPE #	IN-PLACE QUANTITY	UM	TOTAL MANHOURS	TOTAL LABOR COST	INVOICED QUANTITY	UM	TOTAL SUB-PO-MAT COST	ACCOUNTING USE ONLY
Example 1									
99997	MAT	--	--	--	--	(10)	CY	(312.00)	
03310	MAT	--	--	--	--	10	CY	312.00	
Example 2									
99997	LAB	--	--	(80)	(412.00)	--	--	--	
06500	LAB	--	--	80	412.00	--	--	--	
Example 3									
03340	LAB	(50)	CY	--	--	--	--	--	
03360	LAB	50	CY	--	--	--	--	--	
Example 4									
01603	MAT	--	--	--	--	15	LS	--	
01624	LAB	20	LS	--	--	--	--	--	
TOTALS		20		* 0	* 0	15		* 0	

Submitted by: _____ DATE: _____
Processed by(Acct.): _____ DATE: _____
Processed by(DP): _____ DATE: _____

*NOTE: Any adjustments to cost require approval of Division Manager XX
Any transfer of costs between projects require approval of President _____

Figure 9.1. Actual cost and quantity maintenance form.

The first step in correcting this error is to delete the invalid information shown on Line 1 of the first example. The second step is to input the correct information shown on Line 2 of the first example. Similar procedures would be used to make adjustments when invoices have been approved to an incorrect cost code that is valid for the project.

Example 2. Time cards turned in to Accounting indicate labor charges against a cost code that is invalid for the project. A total of 80 labor hours at a total cost of $420 intended for Cost Code 06500, Finish Carpentry, has been charged to Cost Code 65000. This cost code is invalid in the system and is reported on the summary report as Cost Code 99997, Cost Code Error. The same procedures outlined in Example 1 above are used, except that the changes involve labor hours and labor costs.

Similar procedures would be used to make adjustments when labor costs have been reported to an incorrect cost code that is invalid for the project.

Example 3. The in-place quantities reported on the unit completion report are incorrectly added to the wrong cost code. A total of 50 CY of in-place concrete was reported incorrectly to cost code 03340, Concreting Walls. It had been intended for Cost Code 03360, Concreting Columns. The procedures are similar to the above examples, except that only the in-place quantity for each cost code is corrected.

Example 4. This example involves only the updating of quantities that have not been reported on a regular basis. Occasionally accounts where the quantities are being installed or invoiced are so small individually that it is difficult to define except on a cumulative basis. In these instances the updating of the quantities requires only the net increase to each cost code quantity, whether it be in-place or invoiced. For example:

1. Cost Code 01603, Power, is shown on the subcontract, purchase order, and material summary with an invoiced quantity of 20 (20%). The project manager desires to update this after several months of minor invoices to an invoiced quantity of 35 (35%). All that is needed is to show the cost code, document type, and the net increase of 15 (15%) to the invoiced quantity as shown on Line 1 of Example 4.
2. Cost Code 01624, Housekeeping, is shown on the labor summary with an in-place quantity of 75 (75%). The project manager is near the completion of the project and desires to update this to an in-place quantity of 95 (95%). Again, all that is needed is to show the cost code, document type, and the net increase of 20 (20%) to the in-place quantity shown on Line 2 of Example 4.

Each actual cost and quantities maintenance form submitted to Accounting should have the job name, job number, and form number shown. The forms are to be numbered in sequence from the beginning of the project. Forms may not apply to more than one project.

It must be kept in mind that the use of the actual cost and quantities maintenance form is intended to serve only as a tool to make necessary adjustments. Wholesale use of this form to maintain the current status on a project will only result in incorrect cost accounting.

CHAPTER 10

ACCUMULATION OF EXTRAORDINARY COSTS

Unfortunately, in the construction of some projects, it becomes necessary to spend money to undertake an item of work resulting from a lack of performance on the part of a subcontractor or vendor, or to initiate work at the direction of the owner/architect/engineer prior to issuance of a change order, or to perform major corrective work on the project.

In order accurately to accumulate the total cost of each category of work, the following cost code groups are to be used:

1. Cost codes 00001 through 00399—subcontractor/vendor backcharges
2. Cost codes 00400 through 00799—owner/architect/engineer force accounts
3. Cost codes 00800 through 00999—major corrective work

When it becomes necessary on a project to perform work in any of these categories, the project manager should provide the accounting supervisor with project number, cost codes 00001 through 00999, and an appropriate cost code description.

The first cost code for these costs on each project will be the first available number in the respective group.

The cost code description will appear on the appropriate summary report in the cost code applicable to the charges.

As there is nothing estimated for these costs, the cost variance column on the summary reports will reflect the total cumulative amount of the costs as cost overrun.

The specific handling of each category of work is further detailed below.

SUBCONTRACTOR/VENDOR BACKCHARGES

A charge to a subcontractor or vendor for these expenses is defined as a *backcharge*. When it becomes necessary to backcharge a subcontractor or vendor, it is very important that accurate cost accounting of this item of work be established at the outset.

The first step in initiating the backcharge is to notify the subcontractor or vendor officially of the ensuing backcharge. The notification procedure is spelled out in the subcontract agreement. The procedures for creating a backcharge against a purchase order agreement are slightly different from those of the subcontract agreement and are explained in the terms and conditions on the reverse side of the purchase order.

Each cost code number will serve as a holding account for the costs of each anticipated backcharge, until such time as a change order can be issued to the subcontractor or vendor.

No subcontractor or vendor will be invoiced directly for the costs of a backcharge account. A change order is to be issued to cover the cost of the backcharge, plus applicable overhead and profit. Should problems occur in the collection of a billing to the subcontractor or vendor, it directly affects the corporate financial statement.

Every effort should be made to see that all backcharge change orders are issued before the billings (excluding retainage) of the subcontractor or vendor exceed the amount remaining after processing the change order. Processing of a backcharge change order out of the retainage being held after a subcontractor has billed 100% complete causes two problems:

1. By processing an invoice for 100% complete (excluding retainage), the project manager could acknowledge that the work has been satisfactorily performed, and that retainage is being held only because of the payment conditions of the prime contract. The answer to this is usually obtained through legal process, which is both costly and unpredictable.
2. Accounting is required to perform a journal entry to correct the accounts payable on a project when a backcharge is deducted from the retainage being held, regardless of the amount.

Therefore, a backcharge change order prepared to reduce retainage must have "Retainage Reduction Approve" noted in the upper right-hand corner of the change order and be approved by the president prior to being given to accounting.

The backcharge account will never be erased on the summary report. The cost overrun reflected in the backcharge number will be offset by the cost underrun shown in the cost code applicable to the subcontract agreement or purchase order.

OWNER/ARCHITECT/ENGINEER FORCE ACCOUNTS

Initiating work at the direction of the owner, architect, or engineer, prior to issuance of a change order, is to be undertaken only with the approval of the division manager. A complete review of the contract documents should be undertaken to assure that no contractual rights are being waived by proceeding without a change order.

The procedure for initiating and handling of work in the category is identical to that for a backcharge account. Again, it is important to process all change orders with the owner, architect, or engineer prior to billing of the project as 100% complete. Otherwise, it might be alleged that payment of the 100% complete billing will waive any further right to claims by the general contractor.

MAJOR CORRECTIVE WORK

Major corrective work relates to those instances in which the general contractor is required to correct a "major" portion of work, which may or may not be chargeable to the owner or a subcontractor. Generally, any costs that are to be assigned to this cost code are in excess of $5,000 for any given task, and must have the president's approval.

CHAPTER 11

PROJECT CLOSE-OUT PROCEDURES

Eventually every project must be closed out on the books by Accounting. At that time a new project for all "postcompletion costs" is opened with the prefix *C* attached to the closed project number. Also, the historical data base must be prepared. This operation is known as the *Project close-out procedure*. This chapter has three sections: accounting close-out procedures, historical data base, and postwarranty close-out procedures.

ACCOUNTING CLOSE-OUT PROCEDURES

The accounting close-out procedures involve the closing of job records by Accounting, and the establishment of these accounts for all costs expended during the period of warranty required by the contract documents. These accounts are:

1. Cost Code 01920, Accrued Costs to Complete. As explained in more detail below, this account is used to record any accounts payable that remain outstanding at the time of project close-out. Any credits to the project for deposit returns, subcontractor credits, unused materials, and so on should also be applied to this account.

2. Cost Code 01940, Warranty. This is the cost code used for all costs of performing work during the warranty period to correct defective materials and workmanship.

3. Cost Code 01960, Postcompletion Professional Costs. This account, established by the president, is used to record all costs, if any, relating to professional fees during the warranty period of each project.

ROUTINE COURSE OF EVENTS

The account close-out procedures will usually occur in a four-month cycle as established below.

ACCOUNTING CLOSE-OUT PROCEDURES

End of Month 1

When the project status report shows progress billing at 100% of the contract amount and work in process at 95% (or greater) of the total estimated cost, Accounting will issue a memo to the division manager requesting authorization to begin close-out procedures.

End of Month 2

Along with the monthly job summary reports, the project manager will receive a Commitment Run showing the outstanding payables and retainages. (See Figure 11.1).

During Month 3

The project manager should clear up as many accounts as possible and pay all subcontracts, purchase orders, materials, and labor, with the exception of those accounts payable that may be in dispute with the vendor.

End of Month 3

On the invoice cutoff date for Month 3, Accounting will set up the *C* project on the records. After this date, labor costs *only* are to be charged against this new project number (Cost Code 01940, Warranty) until the close-out procedures are completed.

This means that *no* invoices received after the above cutoff date are to be processed until the new project estimate listing (described below) has been entered.

End of Month 4

The final monthly job summary reports and the second commitment run will be distributed. The project manager has the responsibility of preparing the final project status report and completing two other items, the *C* project estimate listing and the project close-out report.

JOB NO.	JOB NAME				COMMITMENT RUN	MONTH OF _____ 19 ____ REPORT IN _____ SEQUENCE			PAGE 00
VENDOR NO.	VENDOR NAME	CONT. NUMBER	RET. %	CONTRACT AMOUNT	BILLED TO DATE	RETAINED TO DATE	PAID TO DATE	BALANCE ON CONTRACT	UNPAID BILLINGS
000000	SAMPLE JOB								

Figure 11.1. Commitment run (printout).

C PROJECT ESTIMATE LISTING

Where the second commitment run indicates that a subcontractor, material vendor, or supplier has not yet billed 100% of the contract amount, the remaining payables will be set aside as an additional anticipated cost to be expended during the warranty period. This additional cost will be included in Cost Code 01920, Accrued Costs to Complete. Given below are two examples of how this additional accrued cost can be handled.

Example 1

XYZ Painting Company has subcontract agreement S0692 with the general contractor in the amount of $100,000. They have billed the project 85% complete through this date. The work in process on this subcontract shows $85,000. Of this amount, the general character is holding a retainage of $8,500.

XYZ Painting Company—contract amount	$100,000
Work in process (WIP)	85,000
Outstanding work in process (WIP)	$ 15,000
Retainage being held	$ 8,500

In order to close out the books on the project, the Work in process (WIP) must be closed on all accounts including this subcontract. The remaining Work in process (WIP) of $15,000 outstanding on the subcontract to XYZ Painting Company, for which they have not billed, should be included in the estimate in Cost Code 01920, Accrued Costs to Complete. The retainage being held of $8,500 remains in the original cost code, to be distributed by Accounting upon approval of a request from the subcontractor.

Example 2

XYZ Painting Company had the same contract amount of $100,000. They have billed 100% of the contract amount. However, through this date, they have not requested the retainage on the project.

XYZ Painting Company—contract amount	$100,000
Work in process (WIP)	100,000
Outstanding work in process (WIP)	0
Retainage being held	$ 10,000

In this example, no money is required to be set aside in Cost Code 01920, Accrued Costs to Complete, as the Work in process (WIP) has been billed in full, and the only amount due the subcontractor is the $10,000 retainage

ACCOUNTING CLOSE-OUT PROCEDURES

being held. This amount will be released by Accounting upon approval from the project manager.

The total approved estimate for all postcompletion costs is to be input on an estimate listing, with the project number being the same as the old number with the prefix *C*. Each outstanding payable should be itemized in this listing for future reference as to accuracy.

The estimate listing is to be prepared itemizing each outstanding payable using the same document type and number as the original source document. Using Example 1 above, the following would be input for this payable:

Cost code = 01920
Document type and number = S0692
Estimated total cost = $15,000

The project manager will *not* enter any estimated costs for warranty into the Cost Accounting System.

All costs incurred for work performed during the warranty period, whether material or labor, should be charged to Cost Code 01940.

All professional fees incurred during the warranty period, and approved by the president, should be charged to Cost Code 01960.

PROJECT CLOSE-OUT REPORT

After reviewing the commitment run, the project manager will be able to complete the project close-out report (see Figure 11.2). This report will reconcile the final project status report with the total work in process as found on the final monthly job summary reports. The mechanics of the report are as follows:

Line 1. A, B, and C—The totals as projected on the project status report

Line 2. A—Total work in process as shown on the final summary reports

Line 2. B and C—Itemized list of outstanding accounts payable per commitment run, as well as other pending material invoices, credits, or backcharges

Line 3. A, B, and C—Warranty estimates to be used for future analysis of warranty period expenditures (do not include in projected final *WIP*)

The project manager will submit the above reports and the estimate listing to the division manager, and subsequently the president, for approval at the same time that ordinary project status reports are due.

PROJECT CLOSE-OUT REPORT

Project Name: _____   Date Prepared: _____

Prepared By: _____   Job Number: _____

1. PROJECTED CONTRACT STATUS
    A. Projected Final Contract Amount $_____
    B. Projected Final Work-In-Process $_____
    C. Projected Final Overhead & Profit (Loss) $_____

2. ANALYSIS OF PROJECTED FINAL W.I.P.
    A. Work-In-Process $_____
    B. Accounts Payable (Itemize) (Use additional sheets if necessary)

       Vendor   Amount
       1. _____ $_____
       2. _____ $_____
       3. _____ $_____
       4. _____ $_____
       5. _____ $_____
       6. _____ $_____
       7. _____ $_____
       8. _____ $_____
       9. _____ $_____
      10. _____ $_____
      11. _____ $_____
      12. _____ $_____
                            Sub-Total Accounts Payable $_____ $_____

    C. Credits or Backcharges
       1. _____ $_____
       2. _____ $_____
       3. _____ $_____
       4. _____ $_____
       5. _____ $_____
                                    Sub-Total Credits $_____ $_____
                TOTAL PROJECTED W.I.P. (Must Equal 1.B. Above) $_____

3. WARRANTY ESTIMATES
    A. Labor $_____
    B. Material $_____
    C. Legal (By President Only) $_____

Figure 11.2. Project close-out report.

ABNORMAL PROJECT CLOSE-OUT PROCEDURES

Occasionally final billing for the project will be submitted, but due to disputes with the owner immediate receipt of final payment will be delayed. Examples of such disputes are debatable work items to be completed, bankruptcy of owner, and so on.

There are certain close-out procedures that are unique for this type of

ACCOUNTING CLOSE-OUT PROCEDURES

project. For example, the time period following final billing in which project close-out begins should be 90 days, which coincides with the time allowed by mechanic's lien laws of most states, for the filing of a lien by the general contractor.

At this point in time, a joint management decision will be made among the president, division manager, and project manager as to the processing of the remaining accounts payable. Thirty days following the date of that decision, the project close-out procedures would be instituted. An example of this situation is shown below:

General contractor's contract with owner (final bill to owner Jan. 1)	$1,000,000
Less net cash received to date	900,000
Outstanding receivables	$ 100,000

If the receivable has not been collected within 90 days, the decision referred to above would be made. For the purpose of this example, the decision is to distribute payables only as required through the mechanic's lien law. Prior to expiration of the general contractor's lien rights under the law, the general contractor should also process its lien against the owner.

With reference to the subcontractors involved in this project, the following breakdown of the payables exists:

Projected total work in process	$ 900,000
Less work in process to date	860,000
Outstanding work in process	$ 40,000
Retainage being held on subcontract	$ 86,000

The billing for the project is still shown on the books as $1,000,000. An outstanding receivable is shown on the books of $100,000.

The outstanding accounts receivable should be handled as a normal aged receivable for the remainder of the fiscal year, or as required by auditing procedures. If the aged receivable has not been collected at that point in time, the president will make a decision whether to continue the accounts receivable as an aged receivable or write it off the books as a bad debt.

The accounts payable of Cost Code 01920, Accrued Costs to Complete, is established as the $40,000 accounts payable that has not been distributed to the subcontractors. The $86,000 of subcontractor retainage is already shown in the Work in process (WIP) of the project, but will be distributed only through the procedures of the mechanic's lien law. It is assumed that the general contractor will perform no additional warranty on this project other than the distribution of the accounts payable as required under the procedures of the mechanic's lien law.

HISTORICAL DATA BASE

The historical data base has been established as a system wherein the cost of completed projects is input onto a data run in cost code sequence in order to furnish information by which future projects may be estimated.

The form upon which the historical data information is to be input will be furnished by Data Processing to the project manager upon request by the president. A copy of the first page of the historical data base form has been included in the appendix. The form is to be filled out, as completely as possible, in cost code sequence.

Those cost codes with particular information of value are to be explained in further detail at the close of the report. An asterisk should be placed on the data form completed by the project manager immediately preceding those cost codes that require further detail. The project manager should then prepare a list of all particulars involving those cost codes and submit it, along with the input data to be keypunched, as part of the historical data base.

Quantities should be included in all areas where possible. This also includes any areas wherein lump sum subcontracts have been written, but the quantities are available. As an example, the masonry is normally taken off by the general contractor, as are such areas as stucco, vinyl flooring, ceramic tile, and so on.

The historical data base is to be completed on all projects within 60 days of final billing. This should be the case for all projects, including those with outstanding receivables and payables. The reason for this is that the outstanding receivables and/or payables that might exist should have no effect upon what the total project cost for historical data purposes should have been.

In preparing the historical data base, the project manager should reconcile the total cost of the project for historical data purposes, as close as possible to the Work in process on the project. Any variance that exists between work in process and the historical data base total should be explained.

The historical data file has been broken down into four major classifications in order to group together the construction projects of similar use. These classifications are:

1. Residential
2. Commercial and institutional
3. Industrial
4. Environmental

Each of the classifications is then further divided into types, and in some classifications types are divided into structures.

POSTWARRANTY CLOSE-OUT PROCEDURE

HISTORICAL DATA FILE CLASSIFICATION SYSTEM

Classification—Residential

Type—1 story	Structure	Example
Type—2 stories		
Type—3 stories		
Type—4 and 5 stories		
Type—6 stories and up		

Classification—Commercial and Industrial

Type—1 story	Structure	Example
Type—2 stories		
Type—3 stories		
Type—4 and 5 stories		
Type—6 stories and up		

Classification—Industrial Construction

Type—1 story and up	Structure	Example

Classification—Environmental

Type—sewer lines	Structure	Example
Type—water lines		
Type—sewage pumping station		
Type—water booster stations		
Type—sewage treatment plants, new		
Type—sewage treatment plants, additions		
Type—water treatment plants, new		
Type—water treatment plants, additions		
Type—sludge handling and/or incineration		

POSTWARRANTY CLOSE-OUT PROCEDURE

The cost codes established for the warranty period will be closed 60 days following the expiration of the warranty period. This will allow sufficient time for the processing of all invoices for last-minute warranty work requests on the project.

 The cost expended on a given project with reference to the warranty will be transferred to the books by Accounting upon the expiration of the warranty period.

CHAPTER 12

COST CODING SYSTEM

DIVISION 1—GENERAL CONDITIONS

01601 THROUGH 01699

Labor

Labor as referred to below shall include unloading, handling, storage and/or installation.

Equipment

Equipment applicable to all cost codes will be included in this division unless indicated otherwise in the description or by the division manager.

01601 **Office and Storage Trailer Rental (LS)**
All costs for office and storage trailers, including rental of warehouse space. Include trailer tie downs. Included here would be any office requirements of the architect or engineer. Moving of these trailers to and from the site would be included in code 01610, Mobilization and Demobilization.

01601.1 **Office and Storage Trailer Rental—Intracompany (LS)**
All costs for office and storage trailers, including rental of warehouse space, which are rented from the company. Include trailer tie-downs. Included here would be any office requirements of the architect or engineer. Moving of these trailers to and from the site would be included in Cost Code 01610, Mobilization and Demobilization.

01602 **Water (LS)**
All costs with reference to construction water. In the Utility Division it will also include the cost of making the service connection and installing the temporary facilities. On the con-

DIVISION 1—GENERAL CONDITIONS 69

ventional projects the service connections should be handled and installed by the plumbing subcontractor with the contractor only receiving a monthly bill from the local water department.

01603 **Power (LS)**
All costs for temporary power and lighting including power company contributions for temporary service. Include power for architect and engineer facilities, if required. Included here is the monthly billing for all single and three-phase power.

01604 **Telephone and Telegraph (LS)**
All costs connected with telephone and telegraph. This will include main office costs according to the telephone log. If you are required to furnish a telephone for the owner's representative, it should be coded in this code.

01605 **Toilets (LS)**
All costs with reference to construction rental toilets or any other facilities for the same purpose located at the job site.

01606 **Drinking Water and Ice (LS)**
All costs for ice, cups, and so on that are necessary for providing drinking water at the job site.

01607 **Supervisory Vehicle Account (LS)**
All costs for gasoline, maintenance insurance, rental, and/or mileage for vehicles used by field supervisory personnel.

01607.1 **Supervisory Vehicle Account—Intracompany (LS)**
All costs for gasoline, maintenance insurance, rental, and/or mileage for vehicles used by field supervisory personnel that are rented from the company.

01608 **Travel and Entertainment (LS)**
All costs for travel and entertainment for people from the main office visiting the jobs, owners, engineers, and so on for projects under contract. It shall also include the costs incurred by field personnel when returning to the main office for business.

01609 **Temporary Roads (SY)**
All costs for installing and maintaining temporary roads. In some cases, this account will be used to apply the extra costs required when the base for the permanent roads is used and regrading and extra base material are required before the permanent road surface construction.

01610 **Construction Yard—Mobilization and Demobilization (LS)**
All costs for mobilization and demobilization of the construction yard facilities or the making of the facilities such as job offices, construction sheds, construction fence, temporary project signs,

and so on. The costs for moving the equipment to and from the job site are coded in the equipment accounts.

Do *not* include material handling in this account. The handling of all materials is to be coded to the applicable account wherein the material is to be used.

This account is to be used only for the setup and removal of the construction yard and equipment and is not intended as a catch-all for other items.

01612 **Winter Protection (LS)**
All costs including fuel that is required for winter construction. Included here would be specific types of equipment required for winter protection, including the rental and maintenance of all heaters and salamanders.

01614 **Storm Protection (LS)**
All costs for the protection of materials or the building while under construction from a major storm. Also included is the clean-up following a major storm. This account is normally an insurable loss account, or in some cases a total loss account, and will not usually be estimated.

01616 **Protection of Completed Construction (LS)**

01624 **Housekeeping (LS)**
All costs relating to cleaning up the project during construction, removal of all trash and debris, cleaning of the construction trailer, makeup, erection, and removal of temporary or construction trash chutes, dump charges, trash truck rental, and gas for the trash truck. Also include rental of trash containers, if applicable.

Materials such as sweeping compounds, solvents, protective window coatings, and so on are included here as used in housekeeping. Included here will be the cleaning and washing of vehicles on the job site, if appropriate.

This account is to be used *only* for housekeeping and is not intended as a catchall for other items.

01625 **Final Cleaning (SF)**
All costs with reference to the final cleaning of the project to satisfy the contract documents, including cleaning of the windows.

01628 **Safety Rails and Barricades (LS)**
All costs for safety rails and barricades, such as balcony rails, buckhoist rails, trash chute rails, shaft holes, handrails, and stairs. If safety rails are procured on a rental basis, the rental should be charged to this account.

DIVISION 1—GENERAL CONDITIONS 71

First aid kits should be charged to code 01645, Tools and Supplies.

01630   **Maintenance of Traffic (LS)**
All costs for traffic control and diversion of traffic required for the safe operations of construction, including vehicular and pedestrian traffic. Included should be costs for setting up and removing barricades, paved crossover lanes, flagmen, rental or purchase of barricades, and so on. Road work to accommodate construction traffic and equipment only should be charged to cost code 01609, Temporary Roads.

01636   **Punch-out (LS)**
This account is to be used for all costs during construction in completion of the final punch list prepared by the architect and engineer. This account is to be used only for the cost incurred with reference to correction of defective materials and workmanship prior to final completion. It is not intended as a catchall for the items of work that have never been performed; those costs are to be charged to the appropriate cost codes wherein the work should have been performed originally.

This account will be closed following completion of the punch list. See Cost Code 01940, Warranty, for all costs during the contractual warranty period.

01640   **Field Office Supplies and Equipment (LS)**
All costs for office supplies in a field office including desks, chairs, copiers, pencils, paper, calculators, and so on.

01645   **Tools and Supplies (LS)**
Material costs for all small tools and supplies. Materials to be incorporated into the completion job are not to be coded to this item. Examples of items to be included here appear on the tool and supply checklist. This will include the purchase or rental of all tools and supplies. Two specific items to be included here are saw service and safety equipment such as first aid kits. Tools and supplies should be individual items costing less than $500. Those items exceeding this amount are in code 01650, Nonmobile Equipment.

01650   **Nonmobile Equipment (LS)**
All costs for the rental or purchase of all nonmobile equipment, except scaffolding (code 01665), used on the project, with the exception of items noted specifically in other cost codes such as Site Work, Dewatering, and so on. Examples of equipment included here are concrete pumps, mortar mixers, masonry saws. For insurance requirements see code 01803, Insurance.

01650.1 **Nonmobile Equipment—Intracompany (LS)**
All costs for the rental from the company of all nonmobile equipment, except scaffolding (code 01665), used on the project, with the exception of items noted specifically in other cost code numbers, such as Site Work, Dewatering, and so on. Examples of equipment included here are concrete pumps, mortar mixers, masonry saws, and so on. For insurance requirements see code 01803, Insurance.

01660 **Mobile Equipment (LS)**
All costs including the rental or purchase of mobile equipment, excluding cranes, used on the project with the exception of those items noted specifically in other cost codes such as site work, Dewatering, and so on. Included here would be front-end loaders, forklifts, and so on. Mobile cranes are to be coded to code 01680, Mobile Cranes—Less Than 50 Tons and code 01685, Mobile Cranes—50 Tons, or More. See code 01803, Insurance.

01660.1 **Mobile Equipment—Intracompany (LS)**
All costs including the rental from the company of mobile equipment, excluding cranes, used on the project with the exception of those items noted specifically in other cost codes such as Site Work, Dewatering, and so on. Included here would be front-end loaders, forklifts, and so on. Mobile cranes are to be coded to code 01680, Mobile Cranes—less than 50 Tons, and Code 01685, Mobile Cranes—50 Tons or More. See code 01803—Insurance.

01665 **Scaffolding (LS)**
All costs for scaffolding. This item should be estimated as required in the divisions and coded in this division, scaffolding here being defined as that used as a working platform. Scaffolding used as formwork is included as a cost of forming.

01670 **Personnel and Equipment Hoists (LS)**
All costs for personnel and equipment hoists. Maintenance of the hoist, including recabling, sandblasting, and painting, would be applied to code 01698, Equipment Maintenance. The power, if an electric-driven machine, would be in code 01603, Power, and fuel, oil and grease would be applied to code 01696, Fuel, Oil, and Grease for gas machines. This would include the cost of new sections, operator's time, erection and dismantling and tie-down bolts, and so on for personnel and equipment hoists. See code 01803, Insurance.

01680 **Mobile Cranes—Less Than 50 Ton (LS)**
Include all costs of any mobile crane less than 50-ton capacity. Included would be any labor if the equipment is rented dry. See code 01803, Insurance.

DIVISION 1—GENERAL CONDITIONS

01680.1 **Mobile Cranes—Less Than 50 Tons—Intracompany (LS)**
Include all costs of renting any mobile crane less than 50-ton capacity from the company. Included would be any labor if the equipment is rented dry. See code 01803, Insurance.

01685 **Mobile Cranes—50 Tons or More (LS)**

01685.1 **Mobile Cranes—50 Tons or More—Intracompany (LS)**

01690 **Tower Cranes (LS)**
All costs applicable to a tower crane including rental, erection, dismantling, operator's time, and climbing labor. Maintenance on tower crane or a service contract would be maintained in code 01698, Equipment Maintenance. Power would be maintained under Code 01603.

01696 **Fuel, Oil, and Grease (LS)**
All costs for gasoline, diesel oil, propane, lubricating, and grease required for equipment that is not production-item allocated. This account only includes items that cannot be allocated to a particular production item and, therefore, would not include any fuel for winter protection, supervisory vehicles, and so on.

01698 **Equipment Maintenance (LS)**
All costs to maintain equipment in the general equipment categories. This item is similar to code 01696, Fuel, Oil, and Grease, and, therefore, would not include costs that can be applied to a particular products account. Include here all labor or operators' time applicable to the equipment maintenance portion.

01700 THROUGH 01799—SUPERVISION

Labor

Labor as referred to below shall include unloading, handling, storage, and/or installation.

Equipment

Equipment applicable to all cost codes will be included in this division unless indicated otherwise in the description or by the division manager.

01705 **Contract Development (LS)**
All costs incurred by the company up until the time a contract is received. Include all contract expenses, such as estimating, travel, phones, lodging, and so on. Include expenses for exploratory equipment such as backhoe to determine water table. Included also in this account is the cost for invitation to bid mailing. This is a precontract cost account.

1710 **Office Supervision (LS)**
Senior project managers, project managers, and assistant project managers should record their time to this code number when they perform any activity with reference to the project. Drafting personnel specifically assigned to a division or particular division manager will use this account to allocate their time to the various projects wherein they are performing work.

Also included in the materials column will be all project-allocated postage, courier service, drafting materials, printing costs, plan costs, bus freight, copying costs, and so on. The material column will, therefore, contain all billings with reference to the processing of shop drawings, plans and specifications, and other incidental drawings required by the office supervision personnel.

01711 **Coordination (LS)**
To include all labor of senior project managers, project managers, and assistant project managers relative to coordination of high-rise and low-rise apartment units. This will include all costs relative to coordination.

01712 **Field Supervision (LS)**
Project superintendents and any assistant superintendents who are assigned to the field will use this code number to account for their time on the various projects. If work is performed in a production item, the time should be coded to that item. Include in materials column the costs of project photographs.

01713 **Field Engineering (LS)**
All costs relative to field engineering and layout of all project elements. Included here would be the labor of a field engineer and any assigned assistants employed on the project. Periodic use of a carpenter, and so on will be coded to the applicable production account.

01714 **Quality Control (LS)**
This account should be used when a contract specifically requires the contractor to furnish a quality control team. This item should include the quality control chief and any assistants required.

01715 **Field Clerks (LS)**
Certain projects may require the services of a field clerk or secretary, either on a full-time or part-time basis. The duties will be to assist the superintendent and/or project manager in whatever manner he or she designates.

01716 **Security Systems and Guard (LS)**
When a guard is required this time shall be coded here. Any production work of this employee or any other employees in the

DIVISION 1—GENERAL CONDITIONS

01700 series should be applied to the applicable code. This item would include security service agencies, dog services, full-time or part-time guard, and so on.

01730 **Personnel Allocation (LS)**
To be used only when the following conditions exist:

1. Inclement weather
2. Holidays
3. No work available
4. Strikes and lockouts
5. Topout parties
6. Severance pay

All worker's compensation claims for medical payment will be processed through the insurance carrier.

This item is not to be intended as a catchall, but rather a location for time so that various cost code numbers and their production rates will not be modified by such things as extensive inclement weather.

01740 **Scheduling (LS)**
To be used by any company personnel preparing and updating progress schedules. This will include the time of both field and office personnel related to scheduling. This item would also include the fees of outside agencies wherein they are used to originate and update schedules.

01780 **Moving Expenses (LS)**
All costs for approved moving of personnel.

01790 **Subsistence (LS)**
All costs for expenses or supplemental rental allowance granted personnel on out-of-town projects.

01800 THROUGH 01899—TAXES AND BONDS

01802 **Labor Tax (LS)**
These are the costs that are applied as a percentage of the direct labor cost. Included are such items as worker's compensation, FICA, unemployment insurance, group insurance, and so on. This cost is applied as a labor item.

NOTE: All cost codes listed below are to be applied as material costs in the system, unless a subcontract or purchase order is written.

01803 **Insurance (LS)**
Builder's risk insurance costs and any other insurance costs required by the contract documents that are over and above the

insurance program of the corporation. Other insurance costs that would be above our normal insurance program would be installation floaters, equipment floaters, railroad protective, blasting, boiler, explosion, and so on. Insurance for automobiles and pickup trucks is in Cost Code 01607.

01804 **Permits and Impact Fees (LS)**
Permit costs would include that of the general contractor and, in areas where applicable, other permit fees may have to be paid by the general contractor; that is, HVAC, electrical, plumbing, curb cuts, tree removal, and so on. Also included in this item number would be impact fees, water meter fees, and sewer tap fees that are to be paid for by the general contractor. This cost code would therefore include all of the governmental agency permits required in order to perform work, except for the occupational license that is a part of the general overhead of the company rather than project overhead.

01805 **General Contract Bonds (LS)**
Surety, sales tax, maintenance bonds required of the general contractor.

01806 **Subcontractor Surety Bonds (LS)**

01807 **Professional Fees (LS)**
These are the fees paid by the company for a particular project wherein the services are required from attorneys, engineers, architects, accountants, drafting services, and so on. These are out-of-house organizations charging the company for services rendered on a project or consulting service. This would also include laboratory or engineering testing or reporting other than for soil and concrete.

01900 THROUGH 01999—POST-COMPLETION COSTS

NOTE: The accounts listed below are to be used only following close out of the project, in accordance with Section 3.10, Closeout Procedures.

01920 **Accrued Costs to Complete (LS)**
This account is to be used for the processing of all accounts payable that remain outstanding at the time of project close-out. This is explained in further detail in Section 3.10, Close-out Procedures.

01940 **Warranty (LS)**
This is a post-construction account for all costs during the contractual warranty period that are incurred with reference to cor-

rection of defective materials and workmanship. This account is to be used only following the close-out of the project. See Cost Code 01636, Punch-out, for costs expended during construction.

01960 **Post-Completion Legal Costs (LS)**
All legal costs incurred during the contractual warranty period are to be charged to this account. Included here would be any costs for expert witnesses, arbitration, court costs, and so on. See Cost Code 01807, Professional Fees, for legal costs expended during construction.

DIVISION 2—SITE WORK

02000 THROUGH 02999

Labor

Labor as referred to below shall include unloading, handling, storage, and/or installation.

Equipment

Equipment applicable to all cost codes will be included in this division unless indicated otherwise in the description or by the division manager.

02000 **Site Development (LS)**
All subcontract costs, including all site work, clearing, excavation, grading, roadways, drainage, paving, and so on to one company, or with division manager approval. When work in items is subcontracted on individual basis use the respective codes.

02010 **Subsurface Exploration (LF)**
All costs, including equipment required for boring, standard penetration tests, core drilling, and so on to obtain subsurface information.

02100 **Clear Site (AC)**
All costs including equipment to clean and grub site, strip and stock topsoil, tree removal or pruning, removal, burying and/or burning of site debris. Also include building removal if the structure is to be saved and moved to another site.

02110 **Demolition (LS)**
All costs, including equipment, to demolish and remove existing structures including existing site concrete work.

02112 **Rework Existing Equipment Foundation (LS)**
All costs to remove, remodel, repair, or replace existing pump bases and other equipment foundation. This code is to be used at the project manager's option in lieu of Code 02110 and applicable Division 3 codes.

02120 **Property Protection (LS)**
All costs for the protection of structures, trees, monuments, piping, roads, walks, and so on to remain on the site or adjacent to the site. Also included are costs for cleanup of existing road, and so on used for hauling. The underpinning of existing structures during construction is in code 02420, Underpinning.

02130 **Erosion Control (SF)**
All costs, including equipment, jute fabric, polyethylene, plastic netting, or other erosion control systems required for site erosion control. Include seeding and sodding that is performed for the express purpose of erosion control.

02210 **Site Grading (CY)**
All costs, including equipment for site cuts, fills, building pads, and rough grading including fill dirt from off site (+ or − 0.1′). Include compaction of fill areas.

02211 **Rock Excavation—Bulk (CY)**
All costs, including equipment, to remove solid rock and boulders in bulk foundations that cannot be handled by standard excavating equipment. This includes the costs of drilling, coring, blasting, and removal of material where required.

02212 **Rock Excavation—Trench (CY)**
All costs, including equipment, to remove solid rock and boulders in trench foundations that cannot be handled by standard excavating equipment.

02220 **Bulk Excavation (CY)**
All costs, including equipment, for the excavation, backfill, and compaction required for mass excavations such as basements, pits, tanks, mat foundations, and so on. Any excavations for footings following this bulk excavation are in code 02222, Foundation Excavation.

02222 **Foundation Excavation (CY)**
All costs, including equipment, for the excavation, backfill, and compaction of the footings, grade beams, pile caps, elevator pits, crane bases, and so on. Include the final grading and compaction of footing bottoms and any compaction performed immediately prior to excavation to keep the banks straight during excavation.

DIVISION 2—SITE WORK

This item does *not* include the bulk excavation performed to reach the top elevation of the foundation (e.g. mass basement excavations wherein foundations are started at an elevation well below grade). See code 02220, Bulk Excavation.

02224 **Trenching (CY)**
All costs including equipment for trenching operations for pipelines, conduits, and so on including backfill and normal compaction requirements.

02225 **Special Fills and Compaction (CY)**
Use this code for all costs expended due to unusual fill requirements such as:

1. Pipes passing over disturbed areas
2. Pipe trenches cut through roadways, walks, and so on
3. Unusually stringent specifications and/or compaction requirements
4. Special fill materials, drainage gravel, sands, and so on

02230 **Exterior Fine Grading (SF)**
Labor, material, equipment, and/or sub, to fine-grade exterior areas for landscaping and drainage. Include respreading of stocked topsoil, hand grading around walks and structures, grading in planters, and so on.

02240 **Soil Stabilization—Pressure Grouting (LS)**
All costs to pressure-grout soils with either cement or chemicals for stabilization purposes.

02242 **Vibroflotation (LS)**

02250 **Soil Poisoning (SF)**
Subprice for termite control.

02275 **Soil Testing (LS)**
All costs for soil testing, including proctors, soil compaction, and soil percolation tests.

02300 **Piling (LF)**
All costs associated with installation of piling of wood, concrete, steel augercast, and so on, including subcontracts for installation. Include any engineering and testing fees for monitoring of installation. Include any labor required for pile cutoffs and removal of excess grout.

02306 **Unloaded Test Piles (LF)**
All costs including equipment for unloading, handling, and driving of unloaded test piling. Included will be cost of cutoffs and disposal of cutoff material and operational labor.

02308 **Pile Load Tests (EA)**
All costs associated with pile load tests including independent laboratory inspection, supervision, and reporting costs.

02310 **Pile Splices (EA)**
All costs related to application of pile splices including reinforcing, drilling, epoxy, forming, and concreting.

02312 **Preformed Pile Holes (EA)**
All costs related to augering and punching through subsurface strata including cost of fabrication or rental of auger or punch. (Cost of crane and crane operator should be charged to applicable Division 1 account.) Augering through embankment to reach natural grade is to be charged to code 02300, Piling.

02314 **Piling Extraction (EA)**
All costs related to extraction and salvage of piling.

02350 **Caissons (LS)**

02410 **Sheeting and Shoring (SF)**
All costs for wood or steel sheeting and sheet piling. Include wales, braces, cribbing, and tieback costs as required. Include removal except for steel sheeting, which is code 02414, Sheet Piling Extraction.

02414 **Sheet Piling Extraction (SF)**
All costs associated with the extraction and salvage of steel sheet piling.

02418 **Pile Jackets (LF)**
All costs related to jacketing piling, including forming, pouring, stripping of forms. Also included should be all costs incurred in preparation of existing piling to receive jacket.

02420 **Underpinning (LS)**
All costs required for the underpinning of foundations on existing buildings. Other protection of existing structures is in code 02120, Property Protection.

02430 **Dewatering (LS)**
All costs including fuel for dewatering of site excavations, including wellpoints, wells, and pumps.

02435 **Dewatering—Minor (LS)**
All costs for minor dewatering of any job. This item shall include all costs for equipment and all materials such as stone for drains, sump materials, and so on for general pumping operations other than wells or wellpoints.

DIVISION 2—SITE WORK

02500 **Storm Drainage Systems (LS)**
Subcontract costs for all types of storm drainage systems including foundation and underslab drainage systems, drainage structures, pipe, culverts, manholes, castings, and so on when issued as a total contract for the storm drainage system.

When the system is built by your forces, then the costs should be broken down into the various items of work including excavation, concrete, formwork, piping, manholes, and so on.

02550 **Site Utilities (LS)**
Subcontract associated with gas, oil, water, steam, and sewer transmission and/or distribution lines including valve and drain pits, water wells, and so on included in a subcontract.

When the system is built by your forces, then the costs should be broken down into the various items of work including excavation, concrete, formwork, piping, manholes, and so on.

02610 **Bituminous Paving (SY)**
Subcontract costs for base, primer, asphalt wearing surfaces, and sealers.

02615 **Concrete Paving (SY)**

02618 **Pavement Marking and Striping (LF)**

02620 **Curb and Gutter (LF)**
Subcontract costs for concrete or asphalt curbs and gutters including excavation, backfill, reinforcing steel, expansion joints, and so on. To be used when a subcontract is issued for the total operation.

When curb and gutter are installed by your forces, it should be estimated in code 03326 for concreting, and code 03126 for formwork.

02630 **Walks (SF)**
Subcontract costs for concrete, asphalt, or other type walks including excavation, grading, reinforcing, expansion joints, and so on. To be used when a subcontract is issued for the total operation.

When walks are installed by your forces it should be estimated in code 03322 for concreting, 03123 for formwork, and 03320 or 03321 for preparation.

02710 **Permanent Fences and Gates (LF)**
All costs for all types of permanent fencing and gates. Temporary construction fencing is included in code 01610.

02720   **Guardrail (LF)**
All costs related to installation of guardrail mounted on posts or on concrete barriers.

02725   **Parking Bumpers (LS)**

02750   **Irrigation Systems (LS)**
Subcontract costs for lawn and site irrigation systems. Water wells are part of code 02550, Site Utilities. This item is to be used when a subcontract is issued for the entire operation. When the system is built by your forces, then the costs should be broken down into the various items of work including excavation, concrete, formwork, piping, manholes, and so on.

02800   **Landscaping—General (LS)**
All costs for landscaping including seeding, sodding, trees, shrubs, mulch, and so on when issued as a total contract for an entire operation. If further breakdown is required use code 02802, Seeding, 02804, Sodding, and 02806, Landscaping—Plants.

02802   **Seeding (SF)**

02804   **Sodding (SF)**

02806   **Landscaping—Plants (LS)**

02910   **Docks (LS)**

02915   **Seawalls (LF)**

02920   **Groins (LF)**

02925   **Sand Cement Riprap (CY)**
All costs incurred in mixing and placing sand cement mixture in bags and placing same on grade. Included are all costs for grouted joint fill, watering, and cleanup following placement. Excavation for toe wall is to be coded to 02222, Foundation Excavation, and preparation of sloped grade is to be coded to 03321, Prepare Grade for Sloped Pavement.

02930   **Rubble Riprap (TN)**
All costs associated with freight, handling, and placement of broken concrete or stone used as riprap.

02935   **Floating Silt Barrier (LF)**
All costs incurred in setting, anchoring, maintaining, and removing floating silt barrier.

02940   **Dredging (LS)**

02945   **Diving (LS)**

02950   **Tennis Courts and Playing Fields (LS)**

02990　**Remodeling Costs (LS)**
The use of this code will be at the option of the division manager, and only with his or her approval.

This account would be used to accumulate all costs for any project so designated. In most instances this would be a remodeling project wherein the use of additional cost codes would serve little useful purpose.

DIVISION 3—CONCRETE

03100 THROUGH 03199—FORMWORK

Labor

Labor as referred to below shall include unloading, handling, storage, and/or installation.

Equipment

Equipment applicable to all cost codes will be estimated here but coded in the 01600 series unless indicated otherwise in the description or by the division manager.

03100　**Formwork Materials (LS)**
All formwork materials may be included in this one code number at the discretion of the division manager. If the division manager does not elect to take this option, formwork should be cost coded by the superintendents to those items of work wherein the formwork was originally ordered. The materials included here would be form plywood, dimensioned lumber, patented panels, ties, hardware, nails, special coatings, expansion materials, and so on for any concrete formwork throughout the entire project.

Dovetail anchor slot and tiles are included in code 04160, Joint Reinforcement, Anchors, and Tie Systems.

03110　**Foundation Formwork (SF)**
All costs to form all foundation formwork with reference to pile caps, grade beams, elevator pit floors and walls, crane bases, continuous footings, and any structural slabs on grade greater than 12 inches in thickness. This is the forming of all foundations and structural slabs on grade, concreted in code 03310, Foundation Concreting.

03118　**Bulkhead Forms (SF)**
All costs for the placing of all bulkheads, either vertical or horizontal.

03120 Form Slabs on Grade (LF)
All costs with reference to the forming of slab edges for first-floor slabs on grade, thresholds, tub drains, buckhoist slabs, up to and including 12 inches in thickness. Also included are any haunches and thresholds poured monolithically with the slab. This will include the original makeup, setting, truing and plumbing, supervision, oiling, stripping, cleaning, and so on. This account is the forming of all slabs concreted in code 03322, Concreting Slabs on grade (6 Inches Thick and Less).

03123 Form Sidewalks (LF)
All costs for the entire forming operation of sidewalk edge forms.

03124 Forming Stairs on Grade (LF)

03125 Depression Forms on Grade (LF)

03126 Form Exterior Curb and Gutter (LF)
All costs to form exterior curb, or curb and gutter, with your labor and materials. The concreting of this item is in code 03326, Concreting Curb and Gutter on grade. If curb and gutter is subcontracted in its entirety, it is in code 02620, Curb and Gutter.

03127 Equipment Bases (SF)

03128 Form for Concrete Encasement and Fill (SF)
All costs for any required formwork for the encasement of pipes and conduit, fill over excavated footings, for mud mats, and for lean fill within structures which were concreted in code 03328, Concrete Encasement and Fills.

03129 Forming Mud Mats

03130 Embeds

03132 Expansion Joints (LF)
All costs for expansion joints, keyhold joints, and so on when they are performed as a separate operation.

03134 Keyways (LF)
All costs for the placing and removal of all keyways.

03136 Water Stops (LF)
All costs for water stops of copper, steel, rubber, PVC, and so on.

03140 Conventionally Formed Straight Walls (SF)
All costs with reference to the forming of all conventionally formed straight walls. This item includes fabrication, installation, and dismantling. Any straight walls that are to be gang-formed are in code 03144, Fabricate Gang-Formed Walls.
Dovetail anchor slot and ties are included in code 04160, Joint Reinforcement, Anchors, and Tie Systems.

DIVISION 3—CONCRETE

03144 **Fabricate Gang-Formed Walls (SF)**
All costs to fabricate and dismantle following construction straight gang-formed wall panels, regardless of the type of wall. Panels are large, reusable units requiring the use of a crane to move and set. The quantity in this cost code is based on the square foot of form area that is to be fabricated.

03145 **Install Gang-Formed Straight Walls (SF)**
All costs to install, strip, and move straight gang wall forms. Included in this item would be the items of material that are involved in the usage of the gang wall forms, such as ties, form coatings, oil, and so on. The fabrication of panels and their dismantling following construction are in code 03144, Fabricate Gang-Formed Walls.

03146 **Conventionally Formed Curved Walls (SF)**
All costs with reference to the forming of conventionally formed curved walls. Any curved walls that are to be gang-formed are in code 03150, Fabricate Curved Gang-Formed Walls.

03150 **Fabricate Curved Gang-Formed Walls (SF)**
All costs to fabricate curved gang-formed panels, regardless of the type of wall. Panels are large, reusable units requiring the use of a crane to move and set. The quantity in this cost code is based on square feet of form area that are to be fabricated.

03151 **Install Gang-Formed Curved Walls (SF)**
All costs to install, strip, and move curved gang wall forms. The material included in this item would be the items of material that are involved in the usage of the gang wall forms, such as ties, form coatings, oil, and so on.

03152 **"Y" Wall Forms (SF)**
All costs with reference to the forming of conventionally formed "Y" walls. This item includes fabrication, installation, and dismantling.

03154 **Slip Form Systems (SF)**
All costs for any slip form concrete forming system.

03160 **Conventionally Formed Columns (SF)**
All costs to make up, install, and dismantle any conventionally formed columns. Include here the forming costs of any round columns. Any columns that are made up in panels and reused extensively will be included in code 03165, Install Column Form Panels.

Dovetail anchor slot and ties are included in code 04160, Joint Reinforcement, Anchors, and Tie Systems.

03164 Fabricate Column Form Panels (SF)
All costs for the makeup and dismantling following construction of column form panels that are to be reused as a unit. The quantity in this cost code is based on square foot of form area that is to be fabricated.

03165 Install Column Form Panels (SF)
All costs for the installation of prefabricated column forms or panel units. The material included in this item would be the items of material that are involved in the usage of the column forms, such as ties, form coatings, oil, and so on.

03170 Conventionally Formed Beams (SF)
All costs to make up, install, and dismantle formwork for all beams. Any beams formed in conjunction with the forming of slabs off grade should be coded to the account for slabs off grade, 03180. Any beam forms that are preassembled into panels and reused more than two times should be included in 03171, Fabricate Beam Form Panels.

03171 Fabricate Beam Form Panels (SF)
All costs for assembling beam form panels that are to be reused as a unit more than twice. Include here the disassembling following construction. The quantity in this cost code is based on square feet of form area that is to be fabricated.

03172 Install Beam Form Panels (SF)
All costs for the installation of preassembled beam form panels, fabricated in 03171.

03180 Forming Conventional Slabs off Grade (SF)
All costs for the forming of any areas of conventional slabs off grade. Also included in this item will be the initial makeup and dismantling of any forms used in this number. Include any beams poured monolithically with the slabs.

03181 Fabricate Fly-Form Panels (SF)
All costs for the makeup and dismantling of fly-form panels at the ground floor level. The lifting of fly-form panels to the second floor after they have been constructed will be in the cost control number for installing fly forms, 03182. The stripping and the dropping to the ground of the fly-form panels from the roof will also be in the number for the installation. This item is for the labor that is performed at the ground-floor level for the construction and dismantling of these panels. Included here is the cost of handling all materials, including moving to and from the job site. Disposal of unsalvageable material is in code 01624, Housekeeping and Cleanup.

DIVISION 3—CONCRETE

The quantity in this cost code is based on the square feet of form area that are to be fabricated.

03182 **Install Fly Forms (SF)**
All costs to install, drop, fly, and reinstall the flying form panels. Also include the vacuuming, oiling, repairing, remaking, reworking, and so on. This item will include the setting of all slab edges, rental of shoring, bolts, ropes, and accessories, and welding time. The original makeup cost of the fly-form panels and their dismantling are included in the code for fabrication of fly forms, 03181. This item includes the cost of filler panels between the fly forms. Any areas of conventional formwork are to be included in code 03180.

Labor only is to be designated on a per-floor basis.

03184 **Pan Slab Forming Systems (SF)**
All costs for any pan slab forming system. Also included in this number would be the rental or purchase of any pan forms, either fiberglass or steel.

03190 **Metal Deck Forming Systems (SF)**
All costs for any patent slab forming systems such as Epicore, Lift-Slab Systems, and so on.

03191 **Precast–Prestressed Forming Systems (SF)**

03192 **Lift Slabs (SF)**

03193 **Forming for Concrete Toppings (LF)**
All costs to install edge forms for concrete topping and wearing surfaces over rough slabs. Reference code 03390, Concrete Topping.

03195 **Forming Stairs off Grade (SF)**
All costs for forming of stairs off grade. Quantity is based on soffit area of stairs and landings.

03198 **Miscellaneous Formwork (SF)**
All costs for miscellaneous formwork items not included elsewhere, with a quantity of less than 100 square feet per item. This will basically be the forming of all items of miscellaneous concreting included in code 03398, Miscellaneous Concreting.

03199 **Rough Hardware (SF)**
All expendable costs for rough hardware used in forming of concrete including nails, snap ties, form oils, pencil rod, and so on. All costs for items that are not expendable are to be included in the cost code where utilized.

03200 THROUGH 03299—REINFORCING

Labor

Labor as referred to below shall include unloading, handling, storage, and/or installation.

Equipment

Equipment applicable to all cost codes will be coded in the 01600 series unless indicated otherwise in the description or by the division manager.

03200 **Reinforcing Steel and Accessories (TN)**
To include all costs for material and/or placement of reinforcing steel and accessories, including the furnishing, fabrication, delivery, unloading, and installation of rebar.
Wherein separate contracts are issued for labor and material, they can be shown separately in the summary reports.

03250 **Masonry Reinforcing (LB)**

03260 **Epoxy-Coated Reinforcing (TN)**

03270 **Stressing Tendons (LB)**
All costs for the installation of stressing tendons and post tension slabs and/or beams to include both the stressing of tendons and cutting off of tails. This item shall also include the labor required to grout the tendons after stressing.

03280 **Welded Wire Fabric (SF)**
All costs for the furnishing and installation of wire mesh in areas other than slab on grade, including such areas as concrete toppings, Epicore, and so on.

03290 **Bar Supports for Reinforcing Steel (LB)**

03300 THROUGH 03399—CONCRETE WORK

Labor

Labor as referred to below shall include unloading, handling, storage, and/or installation.

Equipment

Equipment applicable to all cost codes will be estimated here but coded in the 01600 series unless indicated otherwise in the description or by the division manager.

DIVISION 3—CONCRETE

03300 **Concrete Material (CY)**
The use of this code will be at the option of the division manager and only with his or her approval. This account would be used to accumulate all concrete material costs into one code rather than distributing the concrete amongst the remainder of this series. It should be used, for example, on a project where all concrete is the same psi and small amounts are being placed in various areas, which makes it extremely difficult to allocate the correct quantities. This will usually refer to ready-mix concrete.

If ready-mix concrete is not available, this item shall include all cement, aggregates, and so on, and labor and equipment rental to mix the concrete in the field.

03310 **Foundation Concreting (CY)**
All costs for the placing of all concrete in foundations. This would include wall foundations, column pads, grade beams, pile caps, elevator pits and crane bases, and so on. Include the cost of runways or other preparation and the cleanup of all tools and equipment after the concreting operation is complete.

03320 **Prepare Slabs on Grade (with Wire Mesh) (SF)**
All costs for fine grading, including tractor time, compacting with tamp, placement of polyethelene, and placement of wire mesh for all slabs that include wire mesh. The grading performed at this time is after the plumbers or other trades have completed their work. If this item is to be included in a subcontract with finishing, it should be in code 03330, Concrete Finishing.

03321 **Prepare Slabs on Grade (without Wire Mesh) (SF)**
All costs for fine grading and compacting for slabs on grade that do not include wire mesh.

All costs related to fine grading of embankment area that is to be covered with concrete topping or sand cement riprap.

03322 **Concreting Slabs on Grade (CY)**
All costs for concreting of slab on grade, including walks and steps, patios, ramps, drives, aprons, and building slabs.

03326 **Concreting Curb and Gutter on Grade (CY)**
Labor and concrete material only to pour and finish any concrete curb and gutter. Formwork for this time is included in code 03126. If curb and gutter is subcontracted in its entirety, it is to be coded to 02620.

03327 **Equipment Bases**

03328 **Concrete Encasement and Fills (CY)**
All costs for concrete encasement of pipe, refill due to undercutting, forming fillets, and other lean concrete fill within the structures. Include finishing cost, if required.

03329 **Concrete Mud Mat (CY)**
All costs to place and finish (as required) mud mats.

03330 **Concrete Finishing (SF)**
This item would be used when a subcontract is awarded for the total concrete finishing on a project. Screeds will be included in this item. This code will include items covered in the slab preparation (03321) if one subcontract includes both finishing and preparation. If contracts need to be broken down into the various types of finishes, or if your labor and materials are used, the below listed numbers would be used.

03331 **Nonslip Aggregate Finish (SF)**

03332 **Cure, Protect, and Seal (SF)**
All costs with reference to curing by whatever means are acceptable. Protection shall mean the covering or barricading in such a manner that uncured concrete areas will not be chipped or soiled by construction traffic. Also included would be protection for inclement weather.

03334 **Dry Shake Applications (SF)**

03335 **Saw-Cut Joints (LF)**

03336 **Rubbing (SF)**
All costs for rubbing of exposed areas as required by the engineer's specifications, including specially applied surface materials that are sprayed, rolled, or brush-applied.

03337 **Sandblast Surfacing (SF)**
All costs required to produce a sandblast finish as required by the specifications.

03338 **Cutting and Patching (SF)**
All costs for concrete chipping, grinding, leveling, cutting, patching, and so on. Include chipping excess concrete from around columns, grouting elevator sills, sliding-glass-door tracks, grinding or cutting seams, patching honeycomb, grinding ceilings, and so on. Include here removing snap ties and patching holes where required.

03339 **Repair and Refinish Concrete (SF)**
All costs to remove deteriorated material, that is, to sandblast, patch, and refinish concrete surfaces including Gunite as required.

DIVISION 3—CONCRETE

03340 **Concreting Walls (CY)**
All costs for concreting walls, including shear walls, wing walls, parapet walls, diaphram walls, handrail, and barrier walls. Also included will be any labor for vibrating, signalers, chute tenders, and erection of scaffolding or runways, where applicable.

03360 **Concreting Columns (CY)**
All costs for concreting of columns and pilasters including vibrating labor, signalers, chute tenders, and erection of scaffolding or runways where applicable.

03365 **Concreting Filled Cells (CY)**
All costs for concreting filled cells of brick or block walls. Also included are masonry tie beams and lintels. Include cavity cap material costs where applicable.

03370 **Concreting Beams (CY)**
All costs for concreting all tie beams, lintel beams, and continuous beams that are formed. Also included here are costs for concreting of pile caps. Beams that are poured with the slab off grade should be in code 03380. Include the cost of vibrating labor, signalers, chute attendants, and erection of scaffolding or runways where applicable.

03380 **Concreting Slabs off Grade (CY)**
All costs to concrete slabs off grade, including beams that are poured monolithically with the slab. This item includes the cleaning of bulkheads, vibrating labor, signalers, chute tenders, and erection of scaffolding or runways where applicable.

03390 **Concrete Topping (CY)**
All costs for concrete topping over other slabs. See code 09800, Applied Coatings, for any toppings other than concrete. For forming of this item, see code 03193, Forming for Concrete Topping.

03395 **Concreting Stairs off Grade (CY)**
All cost to place and finish concrete stairs.

03396 **Pneumatically Placed Concrete (LS)**
All costs to place and finish any pneumatically placed concrete such as Gunite or Shotcrete.

03398 **Miscellaneous Concreting (CY)**
All costs to pour and finish various items of miscellaneous concrete, not included elsewhere. All costs for forming of these items are to be included in code 03198, Miscellaneous Formwork.

03399 **Concrete Testing (LS)**
All costs for taking and testing of concrete, including cylinders, slump tests, air entraining, and so on. Also included would be

the costs of any design mixes required by the specifications. If required, the costs of swiss hammer or coring and testing of in-place concrete would be included in this item.

03400 THROUGH 03499—PRECAST

Labor

Labor as referred to below shall include unloading, handling, storage, and/or installation.

Equipment

Equipment applicable to all cost codes will be estimated here but coded in the 01600 series unless indicated otherwise in the description or by the division manager.

03400	**Precast Architectural Concrete (LS)** All costs for the design, fabrication, delivery, and installation of architectural precast components.
03410	**Tilt-up Wall Panels (SF)**
03420	**Precast Structural Members (LS)** All costs for the design, fabrication, delivery, and installation of precast structural members such as columns, walls, and beams.
03430	**Precast Structural Slabs (SF)** All costs for the design, fabrication, delivery, and installation of precast slabs, including prestressed slabs, hollow-core slabs, single- and double-tee slabs.
03440	**Miscellaneous Precast (EA)** All costs for miscellaneous precast items such as benches, planters, sculptures, and so on.

03500 THROUGH 03599—DECKS

Labor

Labor as referred to below shall include unloading, handling, storage, and/or installation.

Equipment

Equipment applicable to all cost codes will be estimated here but coded in the 01600 series unless indicated otherwise in the description or by the division manager.

03500   **Lightweight Insulating Roof Fill (SF)**
         All costs for the furnishing and installation of roof fills of zonolite, vermiculite, perlite, celcore, and so on.
         This item will include the metal decking used to form a composite deck with the roof fill.
         This item must be coordinated with code 05300, Metal Decks, to ensure that the deck is included *one* time.

03510   **Gypsum Concrete Roof Decks (SF)**
         All costs for gypsum concrete roof decks including bulb trees, formboard, wire mesh, vents, and so on.

03520   **Cementitious Wood Fiber Decks (SF)**
         All costs for installation of all cementitious wood fiber deck, such as tectum roof decks, and so on.

DIVISION 4—MASONRY

04000 THROUGH 04999

Labor

Labor as referred to below shall include unloading, handling, storage, and/or installation.

Equipment

Equipment applicable to all cost codes will be estimated here but coded in the 01600 series unless indicated otherwise in the description or by the division manager.

04000   **Masonry Subcontract—Labor and Materials (PC)**
         All material costs for labor and materials, or with the approval of the division manager. This account can be used to accumulate all costs into one number rather than distributing the charges against other numbers in this series.

04010   **Masonry Subcontract—Labor Only (PC)**
         This account is for use when a subcontract for masonry labor only is used. All material purchases are to be distributed among the following numbers in this series.

04100   **Mortar (CF)**
         All costs for mixing mortars for the masonry work. This item would include cement and lime mortars, acid-resisting mortars, premixed mortars, and so on. Additives, admixtures, and coloring agents are also included.

04105 **Masonry Sand (CY)**
All material costs for sand purchased for use in preparing mortar. Labor for central mixing of the mortar and sand would be in code 04100.

04160 **Joint Reinforcement, Anchors, and Tie Systems (LS)**
All material costs for truss-type, ladder-type, wall ties, and so on for joint reinforcing. Labor for installing joint reinforcement would be included in code 04010. Dovetail anchor slots and ties are included in this cost code, but labor to install them is in the appropriate cost code in Division 3, Concrete.

04180 **Control Joints (LF)**
All material costs for all types of vertical control joint material. Labor to install is in Code 04010.

04210 **Brick Masonry Units (PC)**
All costs for all types of brick.

04212 **Brick Pavers (PC)**

04220 **Concrete Unit Masonry (PC)**
All costs for all sizes of concrete masonry units.

04260 **Structural Glazed Units (PC)**

04270 **Glass Masonry Units (PC)**
All costs for all sizes of glass masonry units.

04400 **Stone (SF)**
All costs for marble, limestone, granite, sandstone, cast stones, flagstone, or natural stone veneers in either a rough or cut condition.

04510 **Masonry Cleaning (SF)**

04520 **Masonry Restoration (SF)**
All costs for the removal, replacement, and repair of masonry, masonry joint repair, and the cleaning of old masonry work.

DIVISION 5—METALS

05000 THROUGH 05999

Labor

Labor as referred to below shall include unloading, handling, storage, and/or installation.

DIVISION 5—METALS

Equipment

Equipment applicable to all cost codes will be estimated here but coded in the 01600 series unless indicated otherwise in the description or by the division manager.

05100 **Structural Steel Framing (TN)**
All costs for furnishing both structural metal products as well as erection. If the structural metal products are purchased without erection, they should be coded to material with erection coded to subcontract. The division manager should be consulted if other situations occur such as your own forces doing the erection. See code 09841 for all types of fireproofing.

05200 **Metal Joists (TN)**
All costs for both standard and long-span joists, bridging, anchors, and other accessories when not included in subcontracts or lump sum materials purchased under code 05100.

05300 **Metal Decking (SF)**
All costs for ribbed, fluted, and cellular metal units for floor or roof decks. When used as permanent metal forms or to support composite decking defined by forming systems, the deck would be included in Division 3.

05400 **Light-Gauge Metal Framing (TN)**
All costs for light-gauge structural metal framing and joist systems, metal tubing, fasteners, and accessories.

05500 **Miscellaneous and Ornamental Iron (TN)**
All subcontract costs only for miscellaneous and ornamental iron. On smaller projects, the division manager should be consulted so as to use this account to accumulate all miscellaneous iron, codes 05501 through 05699.

05505 **Structural Steel Door Frames (TN)**
All costs for steel door frames. See code 08100, Hollow Metal Door Frames. An example would be a security door frame fabricated out of miscellaneous shapes.

05510 **Metal Stairs (TN)**
All costs for metal stairs. Concreting of the pans would be in code 03398, Miscellaneous Concreting.

05518 **Ladders (LF)**
All costs for the erection of metal ladders. Include here all costs for the erection of safety cages associated with ladders.

05520 **Wall-Mounted Railings (LF)**

05522 **Floor- or Stringer-Mounted Railings (LF)**
All costs for the erection of floor- or stringer-mounted railings including costs for setting sleeves for railing posts.

05530 **Stair Safety Nosings (LF)**

05535 **Pipe Bumper Posts (EA)**

05540 **Grating Frames and Edge Angles (LF)**
All costs for grating frames, trench cover frames, cover plate frames, dock angles, wall or slab edge angles in formwork, or inserts for precast.

05550 **Gratings (SF)**
All costs for metal grates, including checkered plate trench covers and steel cover plates. See code 15190 for manhole covers and steps.

05560 **Lintels (LF)**

05600 **Miscellaneous Plates and Shapes (TN)**
All costs for the various shapes associated with the work not itemized with the accompanying cost control numbers such as floor plates, miscellaneous angle iron framings for fan openings and accordion door bracing, closure angles, and so on. Steel plates for lining hydraulic rams or compactors would also be included here.

05800 **Structural Expansion Control (LF)**
All costs for expansion joint covers and frames, slide bearings, compression seals, and so on.

DIVISION 6—WOODS AND PLASTICS

06000 THROUGH 06499—ROUGH CARPENTRY

Labor

Labor as referred to below shall include unloading, handling, storage, and/or installation.

Equipment

Equipment applicable to all cost codes will be estimated here but coded in the 01600 series unless indicated otherwise in the description or by the division manager.

DIVISION 6—WOODS AND PLASTICS

06000 **Rough Carpentry (BF)**
All materials may be included in this one item number at the discretion of the division manager. If the division manager does not elect to take this option, rough carpentry material should be coded to the items of work wherein the material was originally ordered.

06010 **Rough Hardware (LS)**
All material costs for nails, hurricane clips, joist hangers, staples, bolts, and so on. This item *does not* include items of trim hardware. Labor to install this material is in the item where installed.

06100 **Partition Framing (Conventional) (BF)**
All costs for rough framing between floors and including all top and bottom plates, studs, jack studs, blocking, dropped ceilings, and fire-stops. This item *does not* include furring.

06108 **Partition Framing (Component Makeup) (BF)**
All costs as described in code 06100 but only as it applies to preassembling partitions or making up tables and placing them in a storage area. Labor for handling and setting these components, once they have left the storage area, would be charged to code 06112.

06112 **Install Preassembled Components (BF)**
All costs for assembly in-place of those items as preassembled under code 06108.

06116 **Post and Beam Framing (BF)**
All costs for *solid* post and beam systems.

06120 **Wood Beams (BF)**
All costs for joist or rafter carriers.

06125 **Floor Joists (BF)**
All costs for floor joists, blocking, bridging, box headers, and plates for joist on masonry or steel. This applies to first-floor joist only. Second-floor joists are considered ceiling joists. See code 06355, Nailers—Lintel Plates.

06128 **Ceiling Joists (BF)**
All costs for ceiling joists and floor joists above first floor, blocking, bridging, box headers, and plates for joist on masonry.

06130 **Rough Stair Framing (BF)**
All costs for the framing of exterior rough stairs or rough framing of interior stairs to be finished later. Include the finishing of exterior stairs in this item number; however, include the finishing of interior rough stairs in code 06570.

Code	Description
06135	**Subflooring—Plywood (BF)**

All costs for plywood sheeting used for subflooring. This item includes sound-absorbing underlayment when laid in conjunction with plywood sheeting.

06140   **Subflooring—T & G and/or Unit Deck (BF)**

06145   **Rafters and Subfascia (BF)**
All costs for rafters, subfascia, collar ties, and blocking.

06155   **Trusses and Subfascia (SF)**
All costs for trusses and subfascia. Square footage is for the projected area covered by trusses.

06160   **Glue-Laminated Construction (BF)**
All costs for glue-laminated beams, including miscellaneous steel plate connections.

06170   **Roof Sheeting—Plywood (BF)**

06175   **Roof Sheeting—Unit Deck (BF)**

06200   **Fender Carpentry (BF)**
All costs of labor and material related to installation of wooden fender systems. (Hardware and cable for installation shall be charged to code 06010, Rough Hardware.)

06250   **Siding—Plywood (BF)**
All costs for exterior plywood siding, trim, shutters, subsiding, and vapor barrier.

06255   **Siding—T & G (BF)**
All costs for exterior T & G siding, trim, shutters, subsiding, and vapor barrier.

06270   **Fascia (LF)**
All costs for finish fascia, drip strips, and so on.

06280   **Soffit (SF)**
All costs for finishing soffit, screening, and soffit molding.

06320   **Furring—Walls (BF)**
All costs for pressure-treated furring strips on walls or other straight runs. The purpose is to differentiate between the high production rates achieved on wall furring and those obtained on furring around doors, windows, and for casework.

06330   **Furring—Doors, Window Bucks and Casework (BF)**
All costs for pressure-treated furring strips around doors and windows, and furring grounds for casework.

06350   **Nailers—Roof Curbs and Parapet Caps (BF) (BF)**
All costs for roof curbs, parapet caps, roofing, and flashing nailers.

DIVISION 6—WOODS AND PLASTICS

06355 **Nailers—Lintel Plates (BF)**
All costs for pressure-treated lintel plates where bolted or attached otherwise to lintels for the purpose of carrying floor joist, ceiling joist, and/or trusses.

06360 **Nailers—Bar Joist Deadwood (BF)**
All costs for deadwood attached to bar joist for anchoring folding partitions, curtain tracks, and so on.

06500 THROUGH 06999—FINISH CARPENTRY/MILLWORK

Labor

Labor as referred to below shall include unloading, handling, storage, and/or installation.

Equipment

Equipment applicable to all cost codes will be estimated here but coded in the 01600 series unless indicated otherwise in the description or by the division manager.

06500 **Finish Carpentry and Millwork (BF)**
All materials may be included in this one cost code at the discretion of the division manager. If the division manager does not elect to take this option, finish carpentry material should be coded to the items of work wherein the material was originally ordered.

06510 **Casework/Millwork (LS)**
All costs for custom-assembled architectural millwork fabricated to field dimensions. Premanufactured cabinetry is to be coded to Equipment in Division 11 wherein the cabinets are to be used.

06550 **Wood Benches and Seats (LF)**
All costs for wood benches or seats including anchoring of pedestals, if any.

06555 **Wood Shelving (SF)**
All costs for wood shelving including closet cleats and closet rods.

06560 **Paneling and Trim (SF)**

06565 **Wood Base (LF)**

06570 **Interior Wood Stairs and Railings—Finished (EA)**
All costs for finishing interior wood stairs and stringer-mounted railings.

06575 **Wood Handrail and Brackets (LF)**
All costs to finish interior-wall-mounted wood handrails and install miscellaneous brackets.

DIVISION 7—THERMAL AND MOISTURE PROTECTION

07000 THROUGH 07999

Labor

Labor as referred to below shall include unloading, handling, storage, and/or installation.

Equipment

Equipment applicable to all cost codes will be estimated here but coded in the 01600 series unless indicated otherwise in the description or by the division manager.

07000 **Thermal and Moisture Protection (LS)**
All subcontract costs for complete thermal and moisture protection or may be used at the option of the division manager to accumulate the costs on certain jobs as broken out by the code numbers below.

07100 **Waterproofing (SF)**
All costs for installation of waterproofing membranes, fluid-applied membranes, liquid membranes, bentonite, metal oxide, and so on.

07110 **Membrane Liners (SY)**
All costs for installation of plastic sheet material used as a waterproof liner for ponds, reservoirs, refuse cells, and so on.

07150 **Damp-Proofing (SF)**
All costs for installation of sealants such as bituminous, silicone, water-repellent coatings, cementitious, and so on. Include here also vapor barriers off-grade (for example, under rigid insulation on concrete decks).

07175 **Caulking (LS)**
All costs for caulking and gaskets where not included in code 07100, Waterproofing. This item to include window, door, and expansion-joint caulking.

07210 **Batt Insulation (SF)**
All costs for batt-type insulation in partition walls and over flat soffit area.

DIVISION 7—THERMAL AND MOISTURE PROTECTION

07211 **Loose Fill Insulation (SF)**
All costs for loose fill insulation in cavities and cells, including soundproofing.

07212 **Rigid Insulation (SF)**
All costs for roof and deck insulation, rigid insulation in cavities, under cooler or freezer floors, glued to precast wall panels, and so on. Where vapor barriers are required when applying rigid insulation over concrete decks, the vapor barrier would be in code 07150.

07213 **Blown Insulation (SF)**

07214 **Foamed-in-Place Insulation (SF)**

07215 **Sprayed-on Insulation (SF)**
All costs for sprayed-on insulation. For sprayed-on fireproofing, see code 09841.

07310 **Shingles (SQ)**
All costs for all types of shingles (wood, asphaltic, etc.) including subfelts. Include here also all flashing and sheet-metal items above the roof line.

07320 **Roofing Tiles (SQ)**
All costs for all types of cement roofing tiles, including subfelts. Include here also all flashing and sheet-metal items above the roof line.

07340 **Membrane Roofing (SQ)**
All costs for membrane roofing such as built-up, prepared rolled roofing, elastic sheet, and fluid applied membranes. Include here also all flashing and sheet-metal items such as gravel stops, above the roof line.

07350 **Traffic Topping (SF)**

07360 **Flashing-below Roof Line (LF)**
All costs for flashing and sheet metal below the roof line including lintel, sill, cavity and cap flashings, and so on.

07365 **Roofing Specialties (LS)**
All costs for such specialties as gutters and downspouts, scuppers, and miscellaneous flashings above the roof line and not included under roofing systems, such as reglets.

07370 **Skylights (EA)**
All costs for skylights. Nailers and roof curbs are in code 06350.

07375 **Hatches (EA)**
All costs for roof hatches. Nailers and roof curbs are in code 06350. Power-type ventilators are in Division 15.

07380 **Gravity Ventilators (EA)**
All costs for gravity-type roof ventilators. Nailers and roof curbs are in code 06350, roof curbs and parapet caps.

07410 **Exposed Roof Panel Systems (SF)**
All costs for preformed roof panels including terne metal roofing.

07450 **Exposed Wall Panel Systems (SF)**
All costs for wall panel systems such as preformed metal siding, composition siding, asbestos cement siding, plastic siding, and so on. Subcontracts for complete metal buildings are in code 13640.

DIVISION 8—DOORS AND WINDOWS

08000 THROUGH 08499—FRAMES AND FINISH HARDWARE

Labor

Labor as referred to below shall include unloading, handling, storage, and/or installation.

Equipment

Equipment applicable to all cost codes will be estimated here, then coded in the 01600 series unless indicated otherwise in the description or by the division manager.

08100 **Hollow Metal Door Frames (EA)**
All costs for hollow metal door frames. In cases where one purchase order is written for all hollow metal, material should be included here with labor cost-coded for each item.

08105 **Hollow Metal Doors (EA)**
All costs for hollow metal doors including installing all hardware. Material only for finish hardware is in code 08400.

08110 **Special Hollow Metal Work (LF)**
All costs for special hollow metal items such as borrowed light openings, transoms, and custom hollow metal work.

08200 **Wood and Plastic Doors and Frames (EA)**
All costs for wood and/or plastic doors and frames including installing necessary hardware. Finish hardware material only is in code 08400. Include here all costs for prehung, pocket, and Dutch doors, including trim.

DIVISION 8—DOORS AND WINDOWS

08300 **Special Doors and Frames (EA)**
All costs for special doors such as Sallyport doors, sliding metal firedoors, metal-clad doors, coiling doors, folding doors, accordion partitions, flexible doors, and so on, including trim. For overhead doors see code 08360.

08360 **Overhead Doors (EA)**

08370 **Aluminum Doors and Frames (EA)**
All costs for aluminum door frames and aluminum doors including installing all finish hardware. Material only for finish hardware is in code 08400.

08400 **Finish Hardware (LS)**
All costs for finish hardware including thresholds, lock sets, door chimes, door bumpers, closers, and so on. Also included would be any locksmith charges. Labor for installation of the hardware would be in codes 08105, 08200, 08300, 08360, and 08370.

08500 THROUGH 08999—WINDOWS, GLASS, AND GLAZING

Labor

Labor as referred to below shall include unloading, handling, storage, and/or installation.

Equipment

Equipment applicable to all cost codes will be estimated here and then coded in the 01600 series unless indicated otherwise in the description or by the division manager.

08500 **Windows, Glass, and Glazing (LS)**
All subcontract cost only for codes 08550 through 08700 when included in a lump sum subcontract. At the option of the division manager, all costs for codes 08500 through 08700 may be accumulated here when the materials cannot be allocated to specific production items.

08550 **Glazing (LS)**
All subcontract costs for glazing. Include types of glass such as plate, sheet, tempered, wire, rough, figured, processed, coated, laminated, and insulating glass, and so on, as well as the glazing plastics and accessories.

08552 **Glass Breakage (LS)**

08570   **Entrances and Storefronts (LF)**

08580   **Operators (EA)**
        All costs for automatic door operators and support equipment.

08585   **Window Walls/Curtain Walls (SF)**

08600   **Sliding Glass Doors (EA)**

08610   **Steel Windows (EA)**

08620   **Aluminum Windows (EA)**

08630   **Wood Windows (EA)**
        All costs for wood windows including window trim.

08650   **Plastic Windows (EA)**

08660   **Special Windows (EA)**
        All costs for special windows not categorized by codes 08610 through 08650.

08700   **Mirrors (SF)**

DIVISION 9—FINISHES

09100 THROUGH 09999

Labor

Labor as referred to below shall include unloading, handling, storage, and/or installation.

Equipment

Equipment applicable to all cost codes will be estimated here and then coded in the 01600 series unless indicated otherwise in the description or by the division manager.

09100   **Lath, Plaster, Stucco Systems (SY)**
        All costs for lath, plaster, and stucco systems. If this account includes lath and scratch coat for ceramic tile, it should be noted in the note of the historical data.

09260   **Gypsum Wallboard Systems (SF)**
        All costs for gypsum wallboard systems, including finishing of wallboard for finish application. This account would include light-gauge metal stud framing, cab backers, rough bucks, if not included in Division 6.

DIVISION 9—FINISHES

09300 **Hard Tile (SF)**
All costs for hard tile surfaces including such tiles as ceramic, ceramic mosaic, quarry marble, glass mosaic, plastic, metal, conductive, and so on.

09400 **Terrazzo (SF)**
All costs for terrazzo, including Portland cement, precast, conductive, plastic matrix, and so on.

09420 **Brown River Gravel Topping (SF)**

09500 **Luminous Ceilings (SF)**

09510 **Acoustical Ceilings (SF)**
All costs for acoustical ceilings, whether lay-in or concealed type, suspended or glue applied.

09520 **Acoustical Wall Treatments (SF)**
All costs for acoustical treatment applied to the wall, not including carpet, which would be in code 09680.

09550 **Wood Floor and Wall Systems (SF)**
All costs for wood flooring systems including such special wall applications as squash court walls. Include such systems as wood strip flooring, parquet, plywood block, resilient wood floor systems for gyms, and so on. Include here also all wood sleeper systems for anchorage.

09555 **Synthetic Gym Flooring (SF)**
All costs for synthetic gym floor systems such as Sport-Tread, Versa-Turf, and so on.

09660 **Resilient Tile Floor (SF)**

09665 **Resilient Sheet Flooring (SF)**

09680 **Carpeting (SY)**
All costs for a complete carpet floor system including carpet cushion.

09800 **Applied Coatings (SF)**
All coats for the application of special coatings, such as epoxy coatings, elastomeric, abrasion resistant, and so on.

09841 **Fireproofing (SF)**
All costs for fireproofing such as fire-resistant coatings, spray-on fireproofing, plaster fireproofing, and so on.

09850 **Aggregate Wall Coatings (SF)**
All costs for aggregate wall coatings such as river gravel set in stucco as a decorative finish.

09900 **Painting (SF)**

09950 **Wall Covering (SY)**
All coats for wall coverings such as vinyl coated, vinyl, cork, wallpaper, wall fabrics, and so on.

DIVISION 10—SPECIALTIES

10000 THROUGH 10999

Labor

Labor as referred to below shall include unloading, handling, storage, and/or installation.

Equipment

Equipment applicable to all cost codes will be estimated here, then coded in the 01600 series unless indicated otherwise in the description or by the division manager.

10100 **Chalkboards and Tackboards (EA)**

10120 **Projection Screens (EA)**

10140 **Trash and Line Chutes (LF)**

10150 **Compartments and Cubicles (EA)**
All costs for compartments and cubicles such as hospital cubicles (curtain tracks), office cubicles, tenant storage compartments, mesh partitions, and so on.

10170 **Fiberglass Shower Stalls (EA)**
All costs for fiberglass shower stalls and tubs when furnished and installed by your own forces. Blocking and partition framing would be in Division 6. If installed under a subcontract, it would be in Division 15.

10180 **Toilet and Shower Partitions (EA)**
All costs for toilet and shower partitions. This account to include laminated plastic toilet partitions, metal toilet partitions, stone partitions, and shower and dressing partitions.

10200 **Louvers and Vents (SF)**
All costs for architectural louvers and vents including bird and insect screens. Louvers for mechanical systems, for example, return air and power operated are in Division 15.

10260 **Wall and Corner Guards (LF)**

DIVISION 10—SPECIALTIES

10270 **Access Flooring (SF)**
All costs for access floor systems, including pedestals, frames, and accessories, such as lifting cups.

10300 **Fireplaces (EA)**
All costs for fireplaces such as prefinished and prefabricated metal units. Include also materials such as ash boxes and dampers when furnished to the mason. All masonry materials such as firebrick and flue liners would be in Division 4. Include here fireplace accessories such as log racks, pokers, and so on.

10350 **Flagpoles (EA)**
All costs for flagpoles. Concrete for the flagpole base should be in Cost Code 03115, Miscellaneous Concreting.

10400 **Interior Identifying Devices (EA)**
All costs for room identifications, direction arrows, and so on. Include also nameplates and valve identifications for mechanical piping when they are not color coded.

10410 **Directories (EA)**

10420 **Plaques (EA)**

10430 **First Aid Cabinets (EA)**
All costs for permanent first aid cabinets called for by the plans. First aid kits for project facilities are in Cost Code 01645, Tools and Supplies.

10440 **Roadway and Exterior Signs (EA)**
All costs for roadway and all other exterior signs such as parking directories, exit and entrance signs, and so on. Parking lot painting would be included in code 02618, Pavement Marking and Striping.

10500 **Lockers (EA)**
All costs for all types of lockers, including miscellaneous anchors necessary for their installation.

10530 **Protective Covers (SF of Coverage)**
All costs for walkway covers and car shelters. This account is intended to encompass such complete covers as aluminum carports, and so on, and not, for example, a structural steel framework.

10550 **Postal Specialties (EA)**
All costs for mail chutes, mailboxes, and so on.

10610 **Demountable Partitions (LF)**

10616 **Movable Gypsum Partitions (LF)**

10650 **Scales (EA)**
All costs for scales, such as truck scales. When purchased as accessories to labs, and so on, they would be coded under code 11865, Laboratory Equipment and Supplies.

10670 **Metal Shelving (SF)**

10750 **Telephone Enclosures (EA)**

10800 **Toilet and Bath Accessories (EA)**
All costs for toilet and bath accessories including medicine cabinets.

10920 **Disappearing Stairs (EA)**

DIVISION 11—EQUIPMENT

11000 THROUGH 11999

Labor

Labor as referred to below shall include unloading, handling, storage, and/or installation.

Equipment

Equipment applicable to all cost codes will be estimated here and then coded in the 01600 series unless indicated otherwise in the description or by the division manager.

11090 **All Anchor Bolts (EA)**
All labor costs to install anchorage for piping, pumps, and equipment. Included are anchor bolts for pumps, generators, blowers engine piping, clarifiers, aerators, filters, conveyors, and so on. If material is not furnished by the equipment supplier and is purchased by general contractor, code to this item.

11091 **Performance Test (LS)**
All costs to start up and test all equipment installed on the project.

11092 **Lubrication (LS)**
All costs for initial lubrication of all equipment. Include also cost of any lubricants required to be supplied for future use. (Some specifications require one year's supply.)

DIVISION 11—EQUIPMENT

11093 **Embeco Grout (CF)**
All costs required for mixing and placing Embeco and similar grouting materials at equipment bases and elsewhere. (Except for grouting, wash water troughs to be coded to code 13149, Wash Water Troughs.)

11100 THROUGH 11199—SOLID WASTE EQUIPMENT

Labor

Labor as referred to below shall include unloading, handling, storage, and/or installation.

Equipment

Equipment applicable to all cost codes will be estimated here and then coded in the 01600 series unless indicated otherwise in the description or by the division manager.

11101 **Shredders, Hammer Mills, and Grinders (EA)**
All costs for shredders, hammer mills, or grinders. Including hoods, bonnets, and shields as required.

11108 **Electric Motors (HP)**
All costs for large-horsepower motors associated with heavy usage in the solid waste equipment.

11112 **Hydraulic Rams (EA)**
All costs for hydraulic rams for garbage compaction. This would include the ramhead cylinders, pumps and lines, and connections to embedded anchors. Steel plate linings should be charged under code 05600, Miscellaneous Plates and Shapes.

11116 **Compactors (EA)**
All costs for compactor equipment.

11128 **Magnetic Separators (EA)**
All costs for magnetic separators, including sprockets, chains, drive motors, and allied structural steel frames.

11200 THROUGH 11499—WATER AND WASTE WATER TREATMENT EQUIPMENT

Labor

Labor as referred to below shall include unloading, handling, storage, and/or installation.

Equipment

Equipment applicable to all cost codes will be estimated here and then coded in the 01600 series unless indicated otherwise in the description or by the division manager.

11300 **Pumps—Material and Unloading (TN)**
All material cost and the labor cost to unload and/or store all types of pumping equipment. (For labor to install, see below.)

11304 **Pumps—Screw Type (EA)**
All labor costs to install and align pumps and drives.

11305 **Pumps—Vertical Close Coupled (HP)**
All labor costs to install and align pumps and drives.

11306 **Pumps—Vertical Extended Shaft (HP)**
All labor costs to install and align pumps and drives.

11307 **Pumps—Vertical Turbine (HP)**
All labor costs to install and align pumps and drives.

11308 **Pumps—Horizontal Centrifugal (HP)**
All labor costs to install and align pumps and drives.

11309 **Pump Drives—Engine (HP)**
All costs for gasoline or diesel engines for pumps. If material is not quoted separate to code 11300, Pumps—Material and Unloading.

11310 **Pump Drives—Miscellaneous Accessories and Trims (LS)**
All labor for miscellaneous accessories and trim. Includes exhaust piping, silencers, radiators, chargers, and so on. Code materials to either 11300 or 11309, as applicable.

11311 **Sump Pumps (EA)**
All labor costs to install and align pumps and drives.

11312 **Pumps—Plunger, Progressive Cavity, Etc. (HP)**
All labor costs to install and align pumps and drives.

11313 **Pneumatic Ejectors and Accessories (LS)**
All costs for pneumatic ejectors and accessories as required.

11314 **Separate Water System—Water Seal Equipment (LS)**
All costs for all equipment as required to provide a separate water seal system.

11315 **Well Systems (LS)**
All costs for well systems including monitoring wells for underground water table and purity. Included here would be cost of drilling and casing wells as required.

DIVISION 11—EQUIPMENT

11316 **Rebuild Existing Pumps and Drives (LS)**
All costs to remove existing pumps and/or drives as required and/or make any repairs or modifications.

11319 **Air Compressors and Accessory Equipment (LS)**
All costs for air compressors and all accessory equipment complete as required.

11320 **Straight-Line Sludge Collectors (TN)**
All costs for sludge collection equipment in rectangular basins.

11321 **Sludge Collector—Scum Troughs (EA)**
Consists of all labor to install scum troughs as required. Material will normally be included and coded in code 11320, Straight-Line Sludge Collectors, or code 11322, Circular Sludge Collectors.

11322 **Circular Sludge Collectors (TN)**
All costs for sludge collection equipment in circular basins.

11323 **Clarifiers—Upflow Type for Water (LS)**
All costs for upflow-type clarifier equipment for water treatment. Welding required would be included under code 11338, Field Welding of Equipment.

11324 **Sludge Thickening Equipment (TN)**
All costs for sludge thickening equipment.

11325 **Grit Collection and Removal Equipment (TN)**
All costs for grit removal, washing, screening equipment, and so on. Air diffusion equipment and air lifts in connection with aerated systems will be coded elsewhere under the appropriate number.

11326 **Flocculators (TN)**
All costs for paddle-type flocculation equipment.

11327 **Sludge Collectors and Flocculators—Adjustable Weirs (LF)**
All costs for adjustable weirs as required. If the material is included in the price of the collectors and flocculators, then code to applicable number.

11328 **Wood or Plastic Baffle (SF)**
All costs for wood or plastic baffles installed in water or sewage treatment plants.

11330 **Digester Equipment—Material and Unloading (TN)**
All material costs and the labor cost to unload and/or store all types of digester equipment. (For labor to install, see below.)

11331	**Digester Equipment—Fabricate Roof Deck (SF)** All labor to fabricate and install roof decks on digesters as required.
11332	**Digester Equipment—Gas Recirculating and Mixing (LS)** All labor to install gas recirculating and mixing equipment for digesters.
11334	**Digester Equipment—Gas Regulating (LS)** All labor to install gas regulating equipment for digesters.
11335	**Digester Equipment—Heat Exchangers (EA)** All labor to install heat exchange equipment for digesters.
11336	**Digester Equipment—Gas Compressors and Accessories (LS)** All labor to install gas compressors with accessories for digesters.
11337	**Floating Digester Equipment—Install (TN)** All labor to install floating digester covers. Welding to be included under code 11338.
11338	**Field Welding of Equipment (LF)** All costs for welding in connection with equipment erection. This includes digester covers, clarifiers, and so on.
11339	**Floating Digester Equipment—Ballast Block (EA)** All costs for casting and installing concrete ballast blocks on floating digester covers.
11340	**Mechanical Mixers and Aerators (EA)** All costs for mechanical digester mixers, aeration mixers, chemical mixers, sludge mixers, and so on.
11341	**Rotary Distributors (TN)** All costs for rotary distributors as used with trickling filter media.
11342	**Aeration Equipment—Material and Unloading (TN)** All material costs and the labor cost to unload and/or store all types of aeration equipment. (For labor to install, see below.)
11343	**Aeration Equipment—Blower and Accessories (EA)** All labor to install blowers and accessories where used for sewage aeration.
11344	**Aeration Equipment—Air Intake Filters (EA)** All labor to install intake filters for aeration blowers.
11345	**Aeration Equipment—Diffuser Assembly (EA)** All labor to install swing and ease-out-type diffuser assemblies in connection with sewage aeration.

DIVISION 11—EQUIPMENT

11346 **Aeration Equipment—Channel Diffuser Assembly (EA)**
All labor to install channel-diffuser-type aerator assemblies.

11347 **Aeration Equipment—Tube Cleaning Tank and Equipment (LS)**
All labor to install aeration tube cleaning tank and accessory equipment.

11348 **Forced Draft Aerators (EA)**
All costs for forced-draft-type aerators complete.

11349 **Airlift Assemblies (EA)**
All costs for airlift assemblies including control valves and control panels.

11362 **Micro Screen Strainers (EA)**
All costs for micro screen strainers.

11363 **Mechanically Cleaned Bar Screens (TN)**
All costs for front or back mechanically cleaned bar screens.

11364 **Comminutors and Barminutors (EA)**
All costs for comminutor equipment including drive motors and extended shafts.

11365 **Chemical Feed Equipment—Materials and Unloading (TN)**
All material costs and the labor cost to unload and/or store all types of chemical feed equipment. (For labor to install, see below.)

11366 **Chlorination Equipment—Scales and Racks (EA)**
All labor to install chlorination weighing systems of any type along with cylinder storage racks.

11367 **Chlorination Equipment—Evaporators (EA)**
All labor to install chlorine evaporators.

11368 **Chlorination Equipment (EA)**
All labor to install chlorinators.

11369 **Chlorination Equipment—Sample Pumps, Analyzers, Etc. (EA)**
All labor to install miscellaneous chlorination equipment. Includes sample pumps, analyzers, leak detection, gas mask cabinets, and so on.

11370 **Chemical Feed Equipment—Feeders (EA)**
All labor to install any type of dry chemical feeder.

11371 **Chemical Feed Equipment—Slakers (EA)**
All labor to install lime slakers.

11372 **Chemical Feed Equipment—Dust Collectors (EA)**
All labor to install dust collectors that mount on chemical feed equipment. This does not include bin-mounted dust collectors.

11373 **Chemical Feed Equipment—Feed Pumps (EA)**
All labor to install all types of chemical feed pumps.

11374 **Ammoniators (EA)**
All labor to install ammoniators used in feeding ammonia gas.

11375 **Ammonia Evaporation Equipment and Accessories (EA)**
All labor to install ammonia evaporators with accessories as required.

11377 **CO_2 Generating Equipment (LS)**
All costs for CO_2 generating equipment completely as required, including interconnecting piping.

11379 **Sludge Filter Presses (TN)**
All material costs for a filter press equipment package. Include under this code the labor to install and assemble the main press. Labor for installing auxiliary items such as tanks, pumps, instruments, and so on shall be coded to the respective item.

11380 **Grease Flotation—Separation Equipment (TN)**
All costs for grease flotation and separation equipment.

11381 **Sludge Dewatering Equipment—Materials and Unloading (TN)**
All material costs and the labor cost to unload and/or store all types of sludge dewatering equipment. (For labor to install, see below.)

11382 **Sludge Dewatering Equipment—Vacuum Filter (EA)**
All labor to install vacuum filter equipment complete with filter cloth or other type of medium.

11383 **Sludge Dewatering Equipment—Conditioning Tank (EA)**
All labor to install sludge conditioning tanks required in connection with vacuum filters.

11384 **Sludge Dewatering Equipment—Vacuum Filtrate Assembly (EA)**
All labor to install vacuum and filtrate assemblies used in connection with vacuum filters.

DIVISION 11—EQUIPMENT

11385 **Sludge Dewatering Equipment—Centrifuge Equipment (EA)**
All labor to install centrifuge equipment where required to dewater sludge.

11386 **Sludge Disposal Vehicle (EA)**
All costs to purchase and equip special vehicles for disposal of sludge from dewatering equipment or sludge drying beds.

11387 **Odor Control Equipment (LS)**
All costs for odor control equipment.

11388 **Sampling Equipment and Accessories (LS)**
All costs to furnish any accessory equipment for sampling purposes on water or sewage.

11389 **Remove Existing Mechanical Process Equipment (LS)**
Consists of all labor to dismantle and/or otherwise remove existing mechanical process equipment. Includes pumps, engines, feeders, chlorine equipment, clarifiers, digesters, metering aerators, filter media, underdrains, wash troughs, conveyors, filters, exchangers, and so on.

11390 **Rain Gauge (EA)**
All costs for rain gauges. If wiring is not in code 16100, include here.

11391 **Instrumentation Equipment—Materials and Unloading (LS)**
All material cost and the labor cost to unload and/or store all types of instrumentation equipment. (For labor to install, see below.)

11392 **Parshall Flume Liners (EA)**
All labor required to set parshall flume liners.

11393 **pH, Dissolved Oxygen Meters, etc. (EA)**
All labor required to install various types of meters such pH, dissolved oxygen, and so on, as required.

11394 **Sludge Density Meters (EA)**
All labor to install sludge density meters.

11395 **All Transmitters (EA)**
All labor to install transmitters as required. This is intended to cover the primary sensing unit, the converter or transducer, and the secondary (sending) unit that then transmits to a remote receiver.

11396 **Remote Instrument Gauges (EA)**
All labor to install receivers as required. This is the unit that

will receive the signal generated by transmitters (code 11395), then convert it as required so the information sent will be in usable form. This code is for individual receivers primarily.

11397 **Instrument and Control Panels (LS)**
All labor to install instrument and/or control panels as required. Many times rather than separate receivers being mounted, they will all be combined into a larger panel. Usually a plant's entire operation is monitored and controlled from a single panel.

11398 **Filter Console (EA)**
All labor to set filter consoles as required. These usually are mounted in front of each filter individually.

11399 **Rebuild and Relocate Existing Metering (EA)**
All labor to rebuild and relocate owner's existing meters.

11400 **Process Control Computers (EA)**
All costs for a complete process control system for a water or sewage treatment plant. This would only be used for a subcontract.

11500 THROUGH 11599—INDUSTRIAL EQUIPMENT

Labor

Labor as referred to below shall include unloading, handling, storage, and/or installation.

Equipment

Equipment applicable to all cost codes will be estimated here, then coded in the 01600 series unless indicated otherwise in the description or by the division manager.

11500 **Industrial Equipment (LS)**
All costs for industrial equipment of a preengineered or preassembled nature, such as concrete batch plants, automated silos, vulcanizing machines, and so on.

11600 THROUGH 11999—SPECIAL EQUIPMENT

Labor

Labor as referred to below shall include unloading, handling, storage, and/or installation.

DIVISION 11—EQUIPMENT

Equipment

Equipment applicable to all cost codes will be estimated here, then coded in the 01600 series unless indicated otherwise in the description or by the division manager.

11610 **Built-In Maintenance Equipment (LS)**
All costs for building maintenance systems such as vacuum systems and power window-washing equipment and related accessories such as hoses, nozzles.

11611 **Mobile Maintenance Equipment (LS)**
All costs for furnishing mobile maintenance equipment and tools such as mowers, pumps, steam cleaners, wrench sets, electric drills, lathes, vacuum cleaners, and so on.

11620 **Bank and Vault Equipment (LS)**
All costs for depository units, safes, teller's windows, teller's counters, vault doors, and so on. See code 14700, Pneumatic Tube Systems.

11630 **Commercial Equipment (LS)**
All costs for equipment for barber shops, beauty salons, retail stores, automobile service stations and other commercial enterprises, including cash registers and display fixtures of all types.

11650 **Checkroom Equipment (LS)**
All costs for checking manual and mechanized equipment such as installed in restaurants and nightclubs.

11660 **Darkroom Equipment (LS)**
All costs for photographic equipment, processors, film and print storage units, developing tanks, print dryers, illuminators, and other specialized equipment including photographic darkroom counter tops and work tables. See code 06510, Casework/Millwork for field-measured millwork, and code 12600, Furniture, for movable accessories.

11680 **Ecclesiastical Equipment (LS)**
All costs for baptismal fonts, chancel fittings, and pews. See Code 12100, Artwork, for other movable accessories.

11700 **Educational Equipment (LS)**
All costs for audiovisual equipment, language laboratories, astro-observatories, and equipment and tools for vocational shops, art, and homemaking, including prefabricated display cabinets used for education purposes.

11740 **Food Service Equipment (LS)**
All costs for commercial and institutional kitchen equipment, tables and machines, cooking equipment, dishwashing equipment, food preparation machines and tables, serving line units, and commercial garbage disposals. See code 11800, Residential Equipment, for packaged residential garbage disposal.

11760 **Bar Units (LS)**
All costs for complete commercial bar units when supplied as package units containing all accessories, railings, stools, etc. See code 06510, Casework/Millwork, for field-measured cabinetry.

11765 **Refrigerated Cases (LS)**
All costs for packaged cases such as used in supermarket displays, commercial kitchens. See code 11800, Residential Equipment, for residential refrigerators. See code 13590, Insulated Room, for nonmobile cold rooms.

11770 **Vending Equipment (LS)**
All costs for package vending machines, coin-operated or not, such as soft drink machines, food vending machines, and so on.

11780 **Athletic Equipment (LS)**
All costs for indoor and outdoor sports equipment, scoreboards, bowling alleys, gym stands, and bleachers (see code 12730, Stadium Seating, for fixed stadium field house seating, or code 12735, Telescoping Bleachers, for telescoping gymnasium seating) and miscellaneous swimming pool equipment when not part of a subcontract pool package. See Division 2, Site Improvements, for playing field and courts.

11790 **Playground Equipment (LS)**
All costs for playground equipment such as swing sets, slides, monkey bars, carousels, and so on.

11800 **Residential Equipment (LS)**
All costs for residential kitchen and laundry appliances and modular residential cabinets when purchased as package units and not field measured. See code 06510, Casework/Millwork, for custom millwork.

11850 **Dock Equipment (LS)**
All costs for leveling platforms, dock levelers, dock bumpers, and so on.

11860 **Portable Ramps, Bridges, and Platforms (LS)**
All costs for airport terminal landside passenger loading equipment.

DIVISION 11—EQUIPMENT

11865 **Laboratory Equipment and Supplies (LS)**
All costs for modular cabinet units, laboratory sinks, tables, fume hoods, instruments, and supplies such as chemical reagents and glassware.

11875 **Laundry Equipment (LS)**
All costs for commercial laundry equipment such as washers, tumblers, extractors, dryers, pressers, and linen trucks, including coin-operated laundry and dry-cleaning equipment. See code 11800, Residential Equipment, for residential laundry appliances.

11880 **Library Equipment (LS)**
All costs for book stacks, charging counters, and related counter work including card catalogs when purchased as modular units.

11900 **Medical Equipment (LS)**
All costs for casework, dental, surgical, and radiological equipment, sterilizers, incubators, and patient care equipment when purchased as modular units. See code 13660, Radiation Protection.

11920 **Mortuary Equipment (LS)**
All costs for embalming tables, mortuary refrigerators, and other mortuary and crematorium equipment.

11925 **Musical Equipment (LS)**
All costs for bells, carillons, organs, and other nonmobile musical instruments.

11930 **Parking Equipment (LS)**
All costs for parking gates, ticket dispensers, and other equipment for parking garages or lots.

11935 **Detention Equipment (LS)**
All costs for cells, cell doors, and other specialized prison equipment such as gun hoppers and security pass-through windows.

11940 **Theater and Stage Equipment (LS)**
All costs for proscenium draperies and curtains, cycloramas, tormentors, flats, demountable risers and platforms, orchestra shells, fly loft rigging and equipment. For public address or commercial projection systems see Division 16, Communications. For stage lighting see Division 16, Lighting. Powered lifts for the installation in stage or orchestra pits are specified in Division 14, Lifts.

11950 **Registration Equipment (LS)**
All costs for units designated for registration, information, and reservations when purchased as modular units.

DIVISION 12—FURNISHINGS

12100 THROUGH 12999

Labor

Labor as referred to below shall include unloading, handling, storage, and/or installation.

Equipment

Equipment applicable to all cost codes will be estimated here, and then coded in the 01600 series unless indicated otherwise in the description or by the division manager.

12100 **Artwork (LS)**
All costs for frescoes, photomurals, ceramics, paintings, sculpture, stained glass, and so on.

12501 **Drapery Rods and Valances (LF)**

12502 **Draperies**
All costs for draperies. See code 11940, Theater and Stage Equipment, for theatrical curtains and draperies.

12504 **Venetian Blinds and Roll Shades (SF)**

12600 **Furniture (LS)**
All costs for free-standing movable furnishings, including desks, tables, and other units appropriate for residential, commercial, and institutional buildings.

12670 **Entrance Mats (EA)**
All costs for entrance mats, such as recessed steel or vinyl mats, rugs, rug cushions, and matting. For carpeting, see Cost Code 09680.

12710 **Auditorium and Theater Seating (EA)**
All costs for permanently affixed seating.

12720 **Restaurant Seating (EA)**

12730 **Stadium Seating (EA)**
All costs for fixed stadium field house or gymnasium seating.

12735 **Telescoping Bleachers (LS)**

12800 **Furnishing Accessories (LS)**
All costs for portable furnishings such as ash and waste receptacles, desk accessories, foliage and fountains, planters, clocks, and so on.

DIVISION 13—SPECIAL CONSTRUCTION

13100 THROUGH 13499—UTILITY CONSTRUCTION

Labor

Labor as referred to below shall include unloading, handling, storage, and/or installation.

Equipment

Equipment applicable to all cost codes will be estimated here, and then coded in the 01600 series unless indicated otherwise in the description or by the division manager.

13101 **Filter Under-Drain—Trickling Filter (SF)**
All costs on filter under-drain for trickling filters in sewage treatment plants.

13110 **Wheeler under Drain Support Piers (CY)**
All costs for support piers for Wheeler type under-drain. Includes formwork, rebar, concrete, and finishing.

13111 **Wheeler Under-Drain—Precast (SF)**
All costs for precast Wheeler under-drain block, grout joints, and installing spheres.

13112 **Wheeler Under-Drain—Cast in Place (SF)**
All costs for formwork, rebar, concrete, finishing, and installing spheres.

13113 **Leopold Under-Drain (SF)**
All costs for vitrified clay tile under-drain as manufactured by F. B. Leopold Company including embedded items and grout.

13114 **Special Under-Drains—Water Plants (SF)**
All costs for any other special under-drain systems (not Wheeler or Leopold).

13120 **Stone Filter Medium—Sewage Plants (TN)**
All costs for crushed stone trickling filter medium used in sewage treatment plants.

13121 **Plastic Filter Medium—Sewage Plants (CY)**
All costs for plastic trickling filter medium used in sewage treatment plants.

13122 **Subgrading Sludge Beds (SF)**
All hand labor required to prepare sludge drying bed subgrade for medium. This is after bulk excavation has been completed.

13123   **Filter Medium—Sludge Drying Beds—Gravel (TN)**
All costs for placing gravel in sludge drying beds.

13124   **Filter Medium—Sludge Drying Beds—Sand (TN)**
All costs for placing sand in sludge drying beds.

13130   **Filter Medium—Water Plant—Gravel (TN)**
All costs for gravel medium to be installed in gravity or pressure filters.

13131   **Filter Medium—Water Plant—Sand (TN)**
All costs for sand medium to be installed in gravity or pressure filters.

13132   **Filter Medium—Water Plant—Anthrafilt (TN)**
All costs for Anthrafilt medium to be installed in gravity or pressure filters.

13133   **Filter Medium—Water Plant—Ferrosand or Zeolite (TN)**
All costs for Ferrosand or Zeolite medium to be installed in gravity or pressure filters or softeners.

13134   **Filter Medium—Water Plant—Level (SF)**
All labor to backwash, scrape, and level filter medium after installation.

13149   **Wash Water Troughs (EA)**
All costs for washing water troughs, necessary leveling, and nonshrink grouting as required.

13150   **Surface Wash Equipment—Surface Washers (EA)**
All costs for the surface washers or "filter sweeps" along with supports and brackets required for same.

13151   **Air Wash Equipment—Piping and Sleeves (EA)**
All costs for piping for air wash equipment as required.

13152   **Air Wash Equipment—Blowers (EA)**
All costs for blower units required in connection with air washing or filters.

13153   **Air Wash Equipment—Filter Screens (SF)**
All costs for filter screens used with air wash equipment.

13154   **Air Wash Equipment—Support Angle and Bars (SF)**
All costs for all support angles and bars for air wash equipment installed in filters.

13211   **Steel Standpipes and Reservoirs (TN)**
This item refers to ground storage-type tanks (those erected at or below ground level). The work will normally be subcontracted.

DIVISION 13—SPECIAL CONSTRUCTION

If not, all material, erection equipment, and erection labor would be coded to this number. Welding would be included under code 13213, Tank Welding.

13212 **Elevated Steel Tanks (LS)**
This item refers to tanks erected on legs above grade. The work will always be subcontracted.

13213 **Tank Welding (Field Fabricated) (LF)**
All material, welding equipment, and labor in connection with tanks erected in the field with your own forces.

13214 **Prestressed Concrete Tanks (LS)**
This item will normally be subcontracted.

13230 **Gas Storage Spheres (LS)**
This item will normally be subcontracted.

13420 **Sludge Incinerators (LS)**
This item will normally be subcontracted.

13430 **Oxygen Generation Equipment (LS)**
All costs for oxygen generation equipment.

13431 **Oxygen Storage Facility (LS)**
All costs for oxygen storage equipment.

13432 **Thermal Sludge Conditioning System (LS)**
This item will normally be subcontracted.

13500 THROUGH 13999—OTHER CONSTRUCTION

LABOR

Labor as referred to below shall include unloading, handling, storage, and/or installation.

Equipment

Equipment applicable to all cost codes will be estimated here, and then coded in the 01600 series unless indicated otherwise in the description or by the division manager.

13550 **Air Supported Structures (LS)**
All costs for pneumatically supported single-and multiple-walled enclosures of plastic, rubber and flexible impregnated materials.

13560 **Audiometric Room (LS)**
All costs for materials and techniques for construction of soundproof rooms for calibration of instruments, testing and research in the field of sound, including structural vibration isolators, sound-absorbing material, sound baffles, and sound retarding doors. Vibration isolating mounts for specific pieces of building equipment should be included with that piece of equipment.

13570 **Clean Room (LS)**
All costs for construction of permanent or demountable rooms designated to establish and maintain a controlled environment free of airborne and other debris of any kind for research and for manufacture of electronic, chemical, pharmaceutical, and other products involving critical tolerances.

13580 **Hyperbaric Room (LS)**
All costs for rooms constructed as large pressure vessels

13590 **Insulated Room (LS)**
To be used solely for complete construction of rooms for cold and frozen storage and for sharp freezing, including insulation, insulated structural panels, insulated doors, floors, shelving, and other equipment and accessories required for a complete installation. Refrigeration equipment may be included in the package but should be noted if it is included here rather than in Division 15, Refrigeration.

13600 **Complete Superstructure (LS)**
To be used when a subcontract is issued for a complete building superstructure to one contractor. Will usually apply to a utility project.

13640 **Prefabricated Buildings (LS)**
All costs for prefabricated buildings, including greenhouses, metal buildings, utility sheds, and so on.

13650 **Screen Enclosures (SF)**
All costs for screen enclosures for apartments, lanais, and pool enclosures. Also included are mechanical equipment screen enclosures. Wire mesh partitions are in Cost Code 10150, Compartments and Cubicles.

13655 **Saunas (LS)**
To be used as a subcontract number only for a complete package sauna. Should the sauna be constructed by your own forces, costs would be broken down, labor and material to their respective code numbers. Be sure that if electrical is not included here it is in Division 16.

13660 **Radiation Protection (LS)**
All costs for radiation shielding materials and their application or incorporation into walls, ceilings, and floors of X-ray and fluoroscopy room, radiological treatment rooms, nuclear accelerator buildings, nuclear reactor installations, fallout shelters, and so on.

13700 **Swimming Pools (LS)**
To be used as a subcontract number only for a complete package swimming pool. If work is done by your own forces, costs would be broken down, labor and material, to their respective code numbers.

DIVISION 14—CONVEYING SYSTEMS

14100 THROUGH 14999

Labor

Labor as referred to below shall include unloading, handling, storage, and/or installation.

Equipment

Equipment applicable to all cost codes will be estimated here, and then coded to the 01600 series unless indicated otherwise in the description or by the division manager.

14100 **Dumbwaiters (EA)**
All costs for manual and power-operated dumbwaiters, including hoistway doors and other related equipment. Will normally be subcontracted.

14200 **Elevators (EA)**
All costs for freight and passenger elevators of all types, including cabs, entrances, controls, safety equipment, tracks and hoistway fittings, and elevator machinery. Will normally be subcontracted.

14300 **Hoists and Cranes (EA)**
All costs for permanent hoisting equipment specified for the project such as overhead trolleys, chain hoists, and so on.

14400 **Lifts (EA)**
All costs for sidewalk lifts, person lifts, platform lifts, stage lifts, ski tows, and so on. Will normally be subcontracted.

14551   **Conveying Systems—Steel Belt (LF)**
All costs for steel conveying equipment. Charge anchor bolts not purchased from the equipment supplier to code 11090, anchor bolts.

14552   **Conveying Systems—Rubber Belt (LF)**
All costs for belt conveyor systems as required. This is for the owner's conveyor that will be a permanent part of the project and is not for contractor's conveyors used for construction.

14553   **Conveying Systems—Pneumatic (LF)**
All costs for pneumatic conveying systems, include entire system complete with piping and valves as required.

14554   **Conveying System—Bucket Elevators (LF)**
All costs for bucket elevators complete. Includes casing and all types of buckets, belt and chair arrangements.

14555   **Conveying Systems—Screw Conveyors (LF)**
All costs for screw conveyors of any type, flight, pitch, casing, trough, and so on, complete. Not for use on feeder screw conveyors that are part of the feeder.

14556   **Conveying Systems—Scales (LS)**
All costs for conveyor scales.

14557   **Hopper and Chute (LS)**
All costs for receiving and discharge hoppers and/or chutes in connection with conveying systems. This covers materials not a part of the conveyor or storage tanks. This would be only materials purchased separately for conveyor system or tanks.

14559   **Bin and Storage Tanks—Day Tanks (EA)**
All costs for day tanks.

14560   **Bin and Storage Tanks—Storage Tanks (EA)**
All costs for large storage tanks used for dry chemicals, grain, solid waste, and so on. Water storage tanks should be coded to code 13211, Steel Standpipes and Reservoirs, code 13212, Elevated Steel Tanks, or code 13214, Prestressed Concrete Tanks. Fuel oil storage tanks should be coded to code 15606.

14561   **Field Welding—Tanks and Bins (LF)**
All costs in connection with welding above tanks erected in the field with your own forces.

14562   **Bin and Storage Tanks—Gauges and Vibrators (EA)**
All costs for all types of liquid or dry level tank gauges, vibrators, and so on.

14563   **Bin and Storage Tanks—Dust Collector (Bin Type) (EA)**
All costs for bin-type dust collectors complete. These are units mounted on the bins.

14570 **Turntables (LS)**
All costs for structural turntables for all applications, movable rooms (such as restaurants, stages, exhibits, and displays) and for vehicular and industrial uses. Will normally be subcontracted.

14600 **Moving Stairs and Walks (LS)**
All costs for passenger conveying systems composed of moving belts or treads installed in horizontal or inclined position complete with handrails, balustrades, tracks, slider plates, motors, controls, safety equipment, and other related items necessary for a complete installation. Will normally be subcontracted.

14700 **Pneumatic Tube Systems (LS)**
All costs for pneumatic tube systems including tubes, fittings, carriers, switching devices, terminal cabinets, air exhausters, controls, and other related equipment. Will normally be subcontracted.

14800 **Powered Scaffolding (LS)**
All costs for scaffolding used primarily as maintenance equipment in a structure. Scaffolding used for construction would be coded to 01665.

DIVISION 15—MECHANICAL

15000 THROUGH 15999

Labor

Labor as referred to below shall include unloading, handling, storage, and/or installation.

Equipment

Equipment applicable to all cost codes will be estimated here and then coded in the 01600 series unless indicated otherwise in the description or by the division manager.

15015 **Setting Inserts—All Pipe Work (LS)**
All costs for inserts in formwork for future supports. Includes all types, even the installation of "read head" anchors for supports after concrete has been poured.

15016 **Hangers, Supports, and Brackets (LS)**
All material costs to install permanent hangers, supports, brackets, and so on, to secure pipe fittings, valves, and so on.

15017   **Cut, Drill, Core, and Grout Holes at Existing or New Structures (LS)**
All material and labor costs to drill or core existing structures and grout holes. This would also include cutting any holes required for any reason in new structures.

15018 THROUGH 15039—CAST IRON AND DUCTILE PIPE LINES

Labor

Labor as referred to below shall include unloading, handling, storage, and/or installation.

Equipment

Equipment applicable to all cost codes will be estimated here and then coded in the 01600 series unless indicated otherwise in the description or by the division manager.

>   NOTE: Codes 15018 through 15032 include labor for all inline valves, flow tubes, and couplings.

15018   **Cast Iron Ductile Iron—Materials and Unloading (TN)**
All material costs and the labor cost to unload and store. Include all joint materials and flexible couplings.

15020   **Cast Iron—Ductile Piping (Manual Erected) through 8" (PC)**
All labor costs to install exposed or buried pipe and fittings that are manually erected with diameter as indicated.

15021   **Cast Iron—Ductile Piping (Manual Erected) 10"–12" (PC)**
Same as code 15020.

15022   **Cast Iron—Ductile Piping (Manual Erected) 14"–18" (PC)**
Same as code 15020.

15023   **Cast Iron—Ductile Piping (Manual Erected) 20"–2405: (PC)**
Same as code 15020.

15024   **Cast Iron—Ductile Piping (Manual Erected) 30"–36" (PC)**
Same as code 15020.

15025   **Cast Iron—Ductile Piping (Manual Erected) 42" and Larger (PC)**
Same as code 15020.

15026   **Cast Iron—Ductile Piping (Machine Erected) through 8" (PC)**
All labor costs to install exposed or buried pipe and fittings that are power machine erected with diameter as indicated.

DIVISION 15—MECHANICAL

15027   **Cast Iron—Ductile Piping (Machine Erected) 10″–12″ (PC)**
Same as code 15026.

15028   **Cast Iron—Ductile Piping (Machine Erected) 14″–18″ (PC)**
Same as code 15026.

15029   **Cast Iron—Ductile Piping (Machine Erected) 20″–24″ (PC)**
Same as code 15026.

15030   **Cast Iron—Ductile Piping (Machine Erected) 30″–36″ (PC)**
Same as code 15026.

15031   **Cast Iron—Ductile Piping (Machine Erected) 42″ and Larger (PC)**
Same as code 15026.

15032   **Cast Iron—Ductile—Steel and Concrete Wall Castings and Sleeves (ID)**
All labor required to install wall pipe, wall sleeves, thimbles, wall castings, and so on, whether poured in place or grouted in later.

15040 THROUGH 15049—CONCRETE AND VITRIFIED CLAY PIPE LINES

Labor

Labor as referred to below shall include unloading, handling, storage, and/or installation.

Equipment

Equipment applicable to all cost codes will be estimated here and then coded in the 01600 series unless indicated otherwise in the description or by the division manager.

> NOTE: Codes 15040 through 15049 include labor for all inline valves, flow tubes, and couplings.

15040   **Concrete and Vitrified Clay Pipe—Materials and Unloading (LS)**
All material costs and the labor cost to unload and/or store pipe and fittings.

15042   **Concrete and Clay Piping—through 18″ (PC)**
All labor costs to install pipe, fittings, and specials with diameter as indicated.

15043   **Concrete and Clay Piping—20″–24″ (PC)**
Same as code 15042.

15044   **Concrete and Clay Piping—30″–36″ (PC)**
Same as code 15042.

15045 **Concrete and Clay Piping—42″–48″ (PC)**
Same as code 15042.

15046 **Concrete and Clay Piping—54″–60″ (PC)**
Same as code 15042.

15047 **Concrete and Clay Piping—60″ and Larger (PC)**
Same as code 15042.

15048 **Concrete Pressure Pipe—Field Weld—all Sizes (ID)**
All costs for field welding.

15049 **Concrete Pressure Pipe—Grouting Joints—All Sizes (ID)**
All costs to grout pipe joints inside and outside.

15050 THROUGH 15062—FABRICATED STEEL PIPE LINES

Labor

Labor as referred to below shall include unloading, handling, storage, and/or installation.

Equipment

Equipment applicable to all cost codes will be estimated here and then coded in the 01600 Series unless indicated otherwise in the description or by the division manager.

> NOTE: All inline items are to be included with the respective pipe codes as is done with cast iron.

15050 **Fabricated Steel Piping—Materials and Unloading (TN)**
All material costs and the labor cost to unload and/or store. Include all joint material and flexible couplings.

15052 **Fabricated Steel Piping—Manually Erected—through 12″ (PC)**
All labor costs to install exposed or buried fabricated steel pipe, fittings, and specials that are manually erected with diameters as indicated.

15053 **Fabricated Steel Piping—Manually Erected—14″–18″ (PC)**
Same as code 15052.

15054 **Fabricated Steel Piping—Manually Erected—20″–24″ (PC)**
Same as code 15052.

15055 **Fabricated Steel Piping—Manually Erected—30″–36″ (PC)**
Same as code 15052.

DIVISION 15—MECHANICAL

15056 **Fabricated Steel Piping—Manually Erected—42″ and Larger (PC)**
Same as code 15052.

15057 **Fabricated Steel Piping—Machine Erected—through 12″ (PC)**
Consists of all labor costs to install exposed or buried fabricated steel pipe, fittings, and specials that must be erected by the use of a machine with diameters as indicated.

15058 **Fabricated Steel Piping—Machine Erected—14″–18″ (PC)**
Same as code 15057.

15059 **Fabricated Steel Piping—Machine Erected—20″–24″ (PC)**
Same as code 15057.

15060 **Fabricated Steel Piping—Machine Erected—30″–36″ (PC)**
Same as code 15057.

15061 **Fabricated Steel Piping—Machine Erected—42″ and Larger (PC)**
Same as code 15057.

15062 **Fabricated Steel Pipe—Field Welds (ID)**
All costs for field welds on steel pipe, whether manual or machine erected.

15063 **Testing—All Piping (LS)**
All costs to pressure-test all lines, cast iron, ductile, steel, concrete, and so on. Includes setup and removal of temporary blocking, plugs, and any other test equipment required.

15064 **Sterilizing All Piping (LS)**
All costs to sterilize all lines, cast iron, ductile, steel, concrete, and so on. Includes setup and removal of any equipment required.

15065 THROUGH 15099—SMALL DIAMETER PIPING AND SPECIALS

Labor

Labor as referred to below shall include unloading, handling, storage, and/or installation.

Equipment

Equipment applicable to all cost codes will be estimated here and then coded in the 01600 series unless indicated otherwise in the description or by the division manager.

15065	**Glass-Lined Pipe and Fittings (PC)** All costs for glass-lined pipe and fittings. Includes, as in part of the material item for this code, any bolts, gaskets, or other joint material required.
15066	**Cast-Iron Soil and Duriron Pipe and Fittings (PC)** All costs for cast-iron soil pipe and fittings or Duriron pipe and fittings, including lead and jute for joints.
15067	**PVC Piping Systems (LS)** All costs for all types of plastic pipe, fittings, valves, specials, and so on, whether exposed or buried. Under the material for this item shall be included all bolts, gaskets, and other joint materials required for this system. Field will report monthly or other designated period percent of completion on this item.
15068	**Fiberglass Piping (LS)** Same as code 15067. Include surface-wash-protected metal pipe under this item.
15069	**Rubber-Lined Piping (LS)** Same as code 15067.
15070	**Saran-Lined Piping (LS)** Same as code 15067.
15071	**Pyrex Piping (LS)** Same as code 15067.
15072	**Stainless Steel Piping (LS)** Same as code 15067.
15073	**Steel—Yoloy Piping (LS)** Same as code 15067.
15074	**Copper Piping (LS)** Same as code 15067.
15075	**Instrumentation Tubing (LS)** All costs for instrumentation tubing, supports, brackets, conduits, trays, bundles, and so on. Tubing may be copper, steel, plastic, aluminum, or stainless steel or as otherwise specified. Field will report periodically percent of completion.
15076	**Hose & Hose Accessories (LS)** All costs for hose and hose accessories.
15080	**Specialties (LS)** All costs for specialties installed on pipe lines, pumps, or equipment, such as pressure gauges, switches, thermometers, nozzles, priming valves, solenoids, release valves, and so on.

DIVISION 15—MECHANICAL

15101 THROUGH 15124—VALVES AND GATES

Labor

Labor as referred to below shall include unloading, handling, storage, and/or installation.

Equipment

Equipment applicable to all cost codes will be estimated here and then coded in the 01600 Series unless indicated otherwise in the description or by the division manager.

15101 **Gate Valves—Material and Unloading (LS)**
All material costs and the labor cost to unload and/or store. Installation labor should be coded to the respective pipe line code.

15102 **Plug Valves—Material and Unloading (LS)**
Same as code 15101.

15103 **Butterfly Valves—Material and Unloading (LS)**
Same as code 15101.

15104 **Ball Valves—Material and Unloading (LS)**
Same as code 15101.

15110 **Check Valves—Material and Unloading (LS)**
Same as code 15101.

15111 **Valve Accessories (EA)**
All labor cost to install valve boxes, extension stems, guides, operators, and so on. Material should be coded to 15101 through 15103.

15112 **Telescoping Valves (EA)**
All costs for telescoping-type sludge and scum draw-off valves, including stem and floorstand.

15115 **Accumulator Equipment (LS)**
All costs for accumulator equipment including hydraulic fluid and control piping to cone or ball valves.

15116 **Mud Valves (ID)**
All costs for mud valves including floor boxes, extension stems, guides, operators, and so on.

15117 **Sluice Gates (ID)**
All costs for sluice and slide gates. Thimble labor costs will be in code 15032, Cast Iron—Ductile—Steel and Concrete Wall Cast-

ings and Sleeves. Labor costs for guides, stems, operators, and so on will be in code 15111, Valve Accessories.

15118   **Fabricated Slide Gates (ID)**
        All costs for fabricated aluminum slide gates and stop gates.

15119   **Float Operated Valves (EA)**
        All costs for float-operated valves.

15120   **Surge Relief Valves (EA)**
        All costs for surge relief valves including field-installed control piping.

15121   **Tapping Sleeves and Valves (EA)**
        All costs for tapping sleeves and valves, complete as specified.

15122   **Inserting Valves (EA)**
        All costs for inserting valves in existing pipe lines.

15123   **Automatic Sewage Regulators (EA)**
        All costs for automatic sewage regulators, complete as specified.

15124   **Hydrostatic Relief Valves (EA)**
        All costs for hydrostatic relief valves.

15125 THROUGH 15187—MISCELLANEOUS PIPING ITEMS

Labor

Labor as referred to below shall include unloading, handling, storage, and/or installation.

Equipment

Equipment applicable to all cost codes will be estimated here and then coded in the 01600 series unless indicated otherwise in the description or by the division manager.

15125   **Chemical Feed Trays (LF)**
        All material and labor costs to install all types of chemical feed trays, complete as specified.

15130   **Pipe Dismantling and Removal (LS)**
        All labor costs to dismantle, remove, haul away, or store in the owner's yard any existing piping systems, whether exposed or buried, including any inline items.

15131   **Pipe Dismantling and Reinstalling (LS)**
        Same as code 15130 except includes cost of reinstalling. Include under this item any joint material that may be required in the reinstallation.

DIVISION 15—MECHANICAL

15185 **Pipe Insulation and Covering (LS)**
All material and labor costs to insulate and cover pipe, fittings, valves, pumps, tanks, equipment, and so on, if performed by your own forces. Will normally be subcontracted.

15186 **Special Coating (LS)**
All material and labor costs to apply special coatings on pipe, fittings, valves, pumps, tanks, equipment, and so on if performed by your forces. Will normally be subcontracted.

15187 **Underground Pipe Insulation (CF)**
All material and labor to install poured-type underground pipe insulation such as Gilsonite. May be subcontracted.

15190 **Manholes (EA)**
All costs to construct manholes other than cast-in-place concrete. To include poured or precast bottoms, brick, concrete block, or concrete pipe sidewall and poured or precast tops. Frames, covers, and steps will be coded to Division 5, Metals, unless cast in precast sections.

15250 THROUGH 15399—PACKAGE TREATMENT FACILITIES

Labor

Labor as referred to below shall include unloading, handling, storage, and/or installation.

Equipment

Equipment applicable to all cost codes will be estimated here and then coded in the 01600 series unless indicated otherwise in the description or by the division manager.

15250 **Package Water Treatment Plant (LS)**
All material and labor to install prefabricated package water treatment plants. Machine rental will normally be included in the 01600 series. If outside rental is required specifically for this item, then code to this number.

15251 **Pressure Filter Equipment (LS)**
All material and labor to install pressure filter equipment including tank, medium, valves, and so on.

15252 **Reverse-Osmosis Modules (EA)**
All material and labor to install reverse-osmosis modules including shells, hose, couplings, and closely associated valves and pipes.

15253   **Degasifiers (EA)**
All material and labor to install degasifiers including core media, filters, filter elements, blowers, and closely associated valves and piping.

15254   **Miscellaneous Skid-Mounted Assemblies (EA)**
All material and labor to install miscellaneous skid-mounted assemblies, such as membrane cleaning systems. Include couplings and closely associated valves, piping, and so on.

15350   **Package Lift Station (LS)**
All material and labor to install prefabricated packaged lift stations as required. Includes connecting piping, valves, precast, wet well, and so on. Machine rental will normally be included in the 01600 series. If outside rental is required specifically for this item, then it should be coded to this number.

15360   **Septic Tank System (LS)**
All material and labor to install precast concrete or steel septic tanks. Drain field tile and medium would be coded to respective items under pipe and medium.

15380   **Package Sewage Plant (LS)**
All material and labor to install package sewage plants as required. Machine rental will normally be included in the 01600 series. If outside rental is required specifically for this item, then code to this number.

15400 THROUGH 15550—PLUMBING AND RELATED ITEMS

Labor

Labor as referred to below shall include unloading, handling, storage, and/or installation.

Equipment

Equipment applicable to all cost codes will be estimated here, then coded in the 01600 series unless indicated otherwise in the description or by the division manager.

15400   **Plumbing and Related Utilities (LS)**
Use this code when plumbing is subcontracted. If breakdown is required, use the following codes for equipment and the applicable piping codes for connection pipe.

15421   **Floor and Shower Drains (EA)**
All material and labor to install floor drains as required. Also includes floor cleanouts if other than a plugged bell type.

DIVISION 15—MECHANICAL

15422 **Roof Drains (EA)**
All material and labor to install roof drains as required. Also includes floor cleanouts if other than a plugged bell type.

15450 **Plumbing Fixtures and Accessories (EA)**
All material and labor to install plumbing fixtures and accessories.

15510 **Sprinkler System (LS)**
Use this code number when sprinkler system is subcontracted.

15515 **Halon Fire Protection System (LS)**

15532 **Fire Hose Cabinets and Accessories (EA)**
All material and labor to install fire hose cabinets and accessories.

15533 **Fire Hose Reels (EA)**
All material and labor to install fire hose reels.

15534 **Fire Hose (LF)**
All material and labor to install fire hose.

15550 **Fire Extinguisher Cabinets and Accessories (EA)**
All material and labor to install fire extinguisher cabinets, including the extinguishers.

15600 THROUGH 15699—HEATING, VENTILATING, AND AIR CONDITIONING

Labor

Labor as referred to below shall include unloading, handling, storage, and/or installation.

Equipment

Equipment applicable to all cost codes will be estimated here and then coded in the 01600 series unless indicated otherwise in the description or by the division manager.

15600 **Heating, Ventilating, and Air Conditioning (LS)**
Use this code number when HVAC system is subcontracted. If breakdown is required, use the following codes for equipment and applicable piping codes for connecting pipe.

15606 **Fuel Oil Tanks and Accessories (EA)**
All material and labor to install fuel oil tanks and accessories as required. Includes all tanks, large and small, above ground and buried, gauges, emergency pumps, and so on.

15620 **Boilers and Accessories (EA)**
All material and labor to install and test boilers as required. Includes accessories and items required for installation, such as ducts.

15700 **Pumps—Feed, Circulating, Etc. (EA)**
All material and labor to install plumbing, heating and cooling related pumps.

15780 **Dehumidifying Equipment—Small Room Type (EA)**
All material and labor to install small-room-type dehumidifiers.

15781 **Dehumidifying Equipment—Large Room Type (EA)**
All material and labor to install large dehumidifying systems.

15810 **Hot Air Furnaces (EA)**
All material and labor to install hot air furnaces.

15829 **Exhaust Fans (EA)**
All material and labor to install ventilating fans and blowers through walls and pedestal mounted.

15830 **Power Roof Ventilators with Curbs (EA)**
All material and labor to install roof exhausters including pre-fabricated curbs.

15840 **Ductwork, Grills, Etc. (LS)**
Will normally be subcontracted. If part of HVAC system, code to 15600.

15880 **Air Conditioning Equipment—Central Types (LS)**
All material and labor to install air conditioning equipment with our own forces. Ductwork to be coded to 15840. If subcontracted, code to 15600.

15881 **Air Conditioning Equipment—Window Type (EA)**
All material and labor to install small package window or through-wall-type air conditioners.

15882 **Air Handling Units (EA)**
All material and labor to install air handling units, unit heaters, and so on with your own forces. If subcontracted, code to 15600.

15900 **Temperature Control Systems (LS)**
Will normally be subcontracted.

15999 **Plumbing, Heating, Ventilating, and Air Conditioning (LS)**
Use this code number when plumbing and HVAC are included in a single subcontract.

DIVISION 16—ELECTRICAL

Labor

Labor as referred to below shall include unloading, handling, storage, and installation.

Equipment

Equipment applicable to all cost codes will be estimated here and then coded in the 01600 series unless indicated otherwise in the description or by the division manager.

16100 **Electrical (LS)**
Use this code number for electrical subcontracts. If breakdown is required, use the following codes.

16210 **Emergency Generator Equipment (LS)**
All cost for emergency generators with diesel or gas-driven engines and the required accessories.

16260 **Automatic Transfer Equipment (LS)**
All costs for automatic transfer switches if not included as part of the package under code 16100 or Code 16210.

16280 **Bridge and Highway Electrical (LS)**
All costs related to installation of road lighting, signalization, navigation lighting, permanent power to equipment, and signaling.

16300 **Power Transmission (LS)**
Use this code number for subcontracts.

16640 **Cathodic Protection (LS)**
Use this code number for subcontracts.

16700 **Communications (LS)**
Use this code number for subcontracts.

16720 **Alarm and Detection Equipment (LS)**

16890 **Electric Heating (LS)**
All costs for electric heating equipment with your own forces. May be subcontracted under code 15600 or code 16100.

GLOSSARY

The definitions that follow are those that it was felt would help most in applying correctly the established procedures of this Cost Accounting System.

In most cases, the definitions apply not to the words used in ordinary day-to-day communication, but to the words in the context of this manual.

It is for this reason that all employees using the Cost Accounting System should have a thorough understanding of each subject, and that this glossary be relied upon for supporting explanations while applying these procedures.

Accounts Payable: An amount paid and/or to be paid on an account. In this system subcontract, purchase order, and materials comprise the accounts payable.

Accounts Receivable: An amount due and/or paid to the company from others on an account.

Average Wage: The average hourly rate of pay within any crew charged against an item of work. (Total labor dollars divided by total man hours.)

Backcharges: Charges to a subcontractor or vendor for work performed by the company, to correct errors or to perform an item of work resulting from a lack of performance on their part. (See Chapter 10.)

Billed to Date: The amount invoiced, including retainage through any specified date. Usually includes completed work plus stored materials.

Change Order: An amendment changing any of the terms on a contract, subcontract, or purchase order. (See Chapter 8.)

Commitment Run: An Accounting output, listing all subcontracts and purchase orders by vendor, and by job, showing adjusted contactor amount, billed to date, retained to date, paid to date, unbilled balance on contract, and unpaid billings at a specified date. (See Chapter 11.)

Contract Document: The signed contract and any change order amendments between the company and an owner.

GLOSSARY

Cost Accounting Committee: A committee appointed by the managing director to assist in the implementing of the Cost Accounting System and to monitor and update same.

Cost Accounting System: A system to record in a concise, readable, and timely form all project costs to date, to forecast project cost at completion and compare both of these with the estimated costs.

Cost, Actual: The amount paid or payable for an item of work.

Cost, Estimated: The amount used in preparing a bid for a project and any subsequent change orders.

Cost, Projected: The anticipated final cost to perform an item of work or complete a project as forecast during construction.

Cost Code: A significant five-digit number assigned to identify an item of work.

Cost Coding Error: Charges made to a cost code that is not valid on a particular project.

Cost Forecasting: The process of projecting from current cost data on any date the final anticipated project cost.

Cost Overrun: The amount the projected final cost exceeds the estimated cost.

Cost to Date: The total actual cost through a specific date.

Cost Underrun: The amount the projected final cost is less than the estimated cost.

Cumulative: The total sum to date of all like elements on a report.

Delivery Ticket: A document received in the field when any shipment is delivered, such as packing slip, freight bill, and so on.

Employee Status Change Form: Form to be used for advising the home office on any change in an employee's status, including classification, salary, and termination. (See appendix.)

Employment Application: Form to be executed by every employee upon initial hiring giving previous employment record, personal information, and starting salary. (See appendix.)

Estimate: The quantity, subcontract cost, material cost, and/or labor cost for each item of work determined at the time of the initial bid on a project, and as adjusted by maintenance forms.

Estimate Balance: This is the difference between the estimate and the cost to date for any item of work.

Estimate Listing: The input to Accounting of the original total project estimate by cost codes, with each cost code separated into quantities, subcontract cost, purchase order material cost, material cost, and labor cost. (See Chapter 1.)

The output from Accounting is the estimate listing printout. (See Figure 1.3.)

Expense Report: Report of personal expenses spent in connection with performance of employee assigned duties. Used to request reimbursement or detail expenditure of an advance. (See Appendix.)

For File Only: A designation placed on the copy of any document being sent to Accounting for information purposes only. (See Chapter 2.)

General Conditions: The items of work on a project related to job overhead as opposed to specific construction of integral parts of the project. Refer to Division 1 of the cost code system.

Historical Data: Cost data on a project retained in the file for use in bidding future projects. (See appendix.)

Inch Diameter: One inch diameter is equal to the product of one unit having a diameter of one inch. As an example, two eight-inch valves would contain sixteen inch diameters. This is a unit of measure used for certain piping and valve items in Chapter 12, Division 15, Mechanical.

On rectangular gates the quantity used per gate will be the average of the length and width (L + W divided by 2.)

Input: Information supplied to Accounting and Data Processing to implement the various system programs.

Item of Work: A specific element of the overall project as broken down by the architect and engineer for specification purposes and by the company for the purpose of recording cost, to which a cost code number has been applied.

Interest Income: Money earned (or lost) as a result of the investment of the net cash flow on a project.

Invoice, Payable: An itemized list from a vendor for materials shipped or from a subcontractor for services performed; specifying the price and terms of payment (See Accounts Receivable.)

Invoices, Receivable: An itemized list prepared by the company, requesting payment from others for materials supplied and/or service rendered. (Also see Accounts Receivable.)

Labor: The services performed by workers for wages on any item of work.

Labor Hour: A unit of one hour's work by one person.

Labor Tax: A percent added to payroll to cover cost of Social Security, unemployment compensation, worker's compensation, health and accident insurance, and payroll courier service. This percentage is reviewed annually and is subject to change.

Lump Sum: A fixed dollar amount to cover the cost of an item of work without regard to the quantity involved.

GLOSSARY

Maintenance Form: This form is used for updating and maintenance of the estimate listing. This form has three uses: input of projected final quantities; input of owner change orders; and redistribution of costs. (See Chapter 8.)

Maintenance Update: Output prepared by Accounting and returned to the project manager following submission of any maintenance form, showing all changes made in the estimate listing. (See appendix.)

Material: The individual elements or substances required in the construction of a project. Used in this system to reference those items for which a purchase order is not issued.

Material Tax: A tax by a government agency for the sales or use of a material within the agency's jurisdiction.

Missing Data: The absence of any input data required by the Data Processing program.

Monthly Transaction Report: Report prepared monthly by Accounting listing by job, vendor, and cost code all subcontract, purchase order material and material accounts paid or payable on each job for the preceding month. (See appendix.)

Output: Information received from Data Processing in connection with the various system programs.

Paid to Date: The total amount paid, excluding retainage, on an account at any specified date.

Production Rates: A quantity of labor required to produce one unit of an item or work (labor hours per unit) or the number of units of an item of work that can be produced by one labor hour (units per labor hour.)

Project Closeout: Closing the books on a project, establishing a warranty account, and preparing the historical data.

Project Cost Code Listing: The listing of the cost codes applicable to a specific project prepared by Data Processing from the Estimate Listing.

Project Facilities: A group of cost codes within Division 1, General Conditions, which are applicable more to job overhead than to a specific item of cost within Divisions 2 through 16. These are further defined in Chapter 12 under codes 01601 through 01698.

Project Number: A significant four-digit number assigned to each project by Accounting for identification purposes.

Project Status Report: A report prepared monthly by the project manager on each project showing adjusted contract amount, project completion date, billings to date, work in process, retainage, projected profit, and other pertinent information summarized from the output documents. (See Chapter 7 and appendix.)

Purchase Order: A document, specifying terms and conditions, issued to a proposed vendor to purchase a certain list of materials. When accepted by the vendor it becomes a source document. (See Chapter 2 and the appendix.)

Purchase Order Material: Materials from a vendor for which a purchase order has been issued and accepted.

Quantity, In-Place: The number of units of an item of work actually installed on a project.

Quantity, Invoiced: The number of units of an item of work delivered and invoiced to the project.

Quantity, Estimated: The number of units of an item of work calculated to be required at the time the original estimate and any subsequent change orders were prepared.

Quantity, Projected Final: The number of units of an item of work the project manager forecasts will be required as the project progresses.

Retainage, Contract: The amount withheld by an owner on each progress billing or invoice.

Retainage, Work in Process: The amount withheld by the company on the invoices submitted by certain subcontractors and suppliers of material and equipment.

Risk Management: The term used to define the technique of analyzing the risks that develop during the progress of construction and allowing for these during the forecasting process.

Source Document: A primary document that supplies information. As used in this Cost Accounting System, it is a copy of a subcontract or purchase order supplied to Accounting along with certain information to input actual vendor and subcontract commitments into the system. (See Chapter 2.)

Statements: A summary of a financial account showing the amount due.

Subcontract: A contract between the company and a third party to provide all or a specified part of the work or materials required on a project. (See appendix.)

Subcontract, Lump Sum: A subcontract under which the subcontractor agrees to complete certain items of work in accordance with plans and specifications on a project regardless of quantity, at an agreed lump sum price.

Subcontract, Unit Price: A subcontract under which the subcontractor agrees to complete a unit of an item of work for an agreed unit price, the final subcontract price being determined by the number of units actually required.

GLOSSARY

Subcontract, Purchase Order, and Materials Summary: Output prepared by Data Processing monthly, which will reproduce the subcontract and material estimate in its entirety, all subcontract and purchase order commitments, and all actual subcontract and material costs to date, and forecast subcontract and material cost at completion with indicated variance from input supplied by the project manager. (See appendix.)

Summary, Check Reimbursement. Summary of nonpayroll checks written in the field with appropriate cost codes assigned and forwarded to Accounting for input to the project cost and reimbursement of the field back account. (See appendix.)

Summary, Monthly Job: This refers to three documents comprising the basic output of the Cost Accounting System. They are the monthly labor summary, the subcontract, purchase order, and materials summary, and the monthly transaction report. (See Chapter 5.)

Summary, Monthly Labor: Output prepared by Data Processing monthly, which will reproduce the labor estimate in its entirety as well as all actual labor costs to date and forecast labor cost at completion with indicated variance from input supplied by the project manager. (See appendix.)

Summary, Petty Cash Reimbursement: Summary of petty cash purchases that are input by Accounting as a project cost on the subcontract, purchase order and materials summary. (See appendix.)

Summary, Weekly Labor: Output prepared by Data Processing weekly, which will reproduce the labor estimate on all items of work having activity during the week and all actual labor costs for the week and to date, and forecast labor cost at completion with indicated variance for these activities from input supplied by the project manager. (See appendix.)

Supervision: A group of cost codes within Division 1, General Conditions, which related more to the management of a project than to the specific items of cost within Division 2 through 16. These are further defined in Chapter 12 in Cost Codes 01705 through 01790.

Time Cards: Preprinted cards prepared by Data Processing for use by the field in reporting and cost coding labor each week. (See appendix.)

Unit Completion Report: Preprinted cards prepared by Data Processing for use by the field and project manager in reporting in place quantities for each cost code each week. (See appendix.)

Unit Cost: The subcontract, material, or labor cost for one unit of an item of work in a project.

Unit of Measure: The commonly accepted unit (i.e., SF, EA, CY, etc.) used for reporting quantity and cost on an item of work. Where no common unit of measure exists the term *lump sum* (LS) is used.

Variance: The difference between the projected final cost of an item of work and the estimated cost. A plus number is a cost overrun. A negative number is a cost underrun. (See Chapter 6.)

Vendor: The seller of a material or product.

Vendor Number: A significant five-digit number assigned to each vendor by Accounting for purpose of identification.

Warranty: The company's responsibility for a specific period to repair or replace any item of defective construction and materials.

Work in Process: This term applies to the total of *all* project costs at any date during the construction period.

Work Week: The starting and ending days of the week established for payroll purposes.

APPENDIX

COST ACCOUNTING FORMS

JOB NAME _____ JOB NUMBER _____ FORM NUMBER _____

COST CODE	DOCUMENT TYPE-#	IN-PLACE QUANTITY	UM	TOTAL MANHOURS	TOTAL LABOR COST	INVOICED QUANTITY	UM	TOTAL SUB-PO-MAT COST	ACCOUNTING USE ONLY
TOTALS					*			*	

Submitted by: _____ DATE: _____
Processed by (Acct.): _____ DATE: _____
Processed by (DP): _____ DATE: _____

*NOTE: Any adjustments to cost require approval of Division Manager.
Any transfer of costs between projects require approval of President _____

Actual cost and quantity maintenance form.

					COMMITMENT RUN	MONTH OF REPORT IN _____ 19___ SEQUENCE		PAGE 00
JOB NO.	JOB NAME							
VENDOR NO.	VENDOR NAME	CONT. NUMBER	RET. %	CONTRACT AMOUNT	BILLED TO DATE	RETAINED TO DATE	PAID TO DATE	BALANCE ON CONTRACT	UNPAID BILLINGS
000000	SAMPLE JOB							

Commitment run.

NAME _____

DATE EFF. __/__/__ HOURLY RATE _____

JOB TITLE _____

OTHER INFORMATION _____

_____
_____
_____
_____
_____
_____
_____
_____
_____
_____
_____
_____
_____

_____
Authorized Signature

Employee status change.

EMPLOYEE APPLICATION

This form has been specially designed to assist you. Please answer all questions carefully. This form will be retained as a permanent record if you are employed. All information will be kept confidential.

PERSONAL INFORMATION

POSITION APPLYING FOR _____ DATE _____
NAME _____ SOCIAL SECURITY NO. _____
STREET ADDRESS _____
CITY, STATE, ZIP CODE _____
PHONE NO. _____ BIRTHDATE _/_/_ AGE ___ MARITAL STATUS: SINGLE___ MARRIED___
SPOUSE'S NAME _____ NO. OF DEPENDENTS _____
HAVE YOU EVER BEEN EMPLOYED BY OUR COMPANY BEFORE?   YES ___ NO ___
IF SO, EXPLAIN _____
ARE ANY OF YOUR RELATIVES PRESENTLY EMPLOYED HERE?   YES ___ NO ___
IF SO, EXPLAIN _____
IN CASE OF ACCIDENT PLEASE CONTACT:
NAME _____ ADDRESS _____ PHONE _____
NAME _____ ADDRESS _____ PHONE _____

EMPLOYMENT RECORD

Please be accurate and give full and complete record of your employment. Begin with present or last position. Use additional sheet of paper if space provided is inadequate.

EMPLOYER		
ADDRESS		
DATES OF EMPLOYMENT		
JOB TITLE		
IMMEDIATE SUPERVISOR	PRESENT OR LAST POSITION	PREVIOUS POSITION
EMPLOYER		
ADDRESS		
DATES OF EMPLOYMENT		
JOB TITLE		
IMMEDIATE SUPERVISOR	2nd PREVIOUS POSITION	3rd PREVIOUS POSITION

ARE YOU PRESENTLY ENROLLED OR HAVE YOU EVER BEEN ENROLLED IN A REGISTERED APPRENTICESHIP PROGRAM?
YES ___ NO ___ IF SO, EXPLAIN _____

Employee application.

Employee Application (*Continued*)

PHYSICAL AND MEDICAL RECORD

DO YOU HAVE ANY PHYSICAL OR MENTAL DISABILITIES OR IMPAIRMENTS? YES ___ NO ___
IF SO, EXPLAIN ___
Do you have or have you ever had any of the following physical conditions, ailments, or diseases? Answer YES or NO.
If YES, give details as to time, duration, treatment and names of Doctors below.

	YES	NO		YES	NO		YES	NO
Tuberculosis			Rheumatism or Gout			High Blood Pressure		
Silicosis			Rupture or Hernia			Heart Trouble		
Asthma			Back Trouble/Injury			Varicose Veins		
Lead Poisoning			Neck Trouble/Injury			Arteriosclerosis		
Diabetes			Osteomyelitis			Hearing Defects		
Epilepsy			Dermatitis			Eye Trouble		
Arthritis			Allergy			Wear Glasses		

HAVE YOU BEEN IN THE ARMED FORCES? YES ___ NO ___ ARE YOU RECEIVING DISABILITY BENEFITS? YES ___ NO ___
HAVE YOU EVER HAD A JOB-CONNECTED DISEASE OR INJURY? YES ___ NO ___ IF YES, DID YOU RECEIVE
ANY COMPENSATION? YES ___ NO ___ MEDICAL BENEFITS? YES ___ NO ___ FOR HOW LONG? ___
HOW MANY TIMES IN THE PAST ONE YEAR HAVE YOU BEEN UNABLE TO WORK BECAUSE OF ILLNESS? ___
EXPLAIN ___

I hereby certify that all statements made on this application are true and correct to the best of my knowledge and belief. I understand that if I have misrepresented myself, I will be subject to immediate dismissal, and I will hold the company harmless for all circumstances relating to any misrepresentation. The latter will include holding the Company harmless for any pre-existing physical conditions.

_____   _____
Date   Signature

FOR OFFICE USE ONLY

REMARKS ___
HOURLY RATE ___ POSITION ___ DATE OF HIRE ___
EMPLOYEE CODE ___ APPROVED BY ___

ESTIMATE LISTING & MAINTENANCE FORM

JOB NAME_____ JOB NO._____ MAINTENANCE FORM NO._____

COST CODE	DOCUMENT TYPE #	ESTIMATED QUANTITY	U/M	ESTIMATED TOTAL MH	ESTIMATED TOTAL COST	PROJECTED QUANTITY
TOTALS						

Submitted by:_____ Date:_____
Processed by (Accounting):_____ Date:_____   PAGE ____ OF ____
Processed by (Data Process.):_____ Date:_____

Estimate listing and maintenance form.

```
ESTIMATE LISTING                                                              DATE RUN  0/00/00                           PAGE 00
DIV._____                    PROJ. MGR._____              JOB NO. & NAME 0000   SAMPLE JOB

*********************************************************************************************************************************
COST CODE      UM   *                       C U R R E N T   E S T I M A T E   D A T A
-------------- ---- *                                                                              EST        ESTIMATED  PRODUCTIVITY
CODE CODE DESCRIPTION    EST. QTY.  EST MAN HRS  EST TOT COST  PROJ. QTY.  EST UNIT COST  AVG WG      U/MH         MH/U
*********************************************************************************************************************************
00000            UM
SAMPLE COST CODE
SUB.                        .00                       .00                     .00
P.O.                        .00                       .00                     .00
MAT.                        .00                       .00                     .00          .00       .00         .00
LAB.                        .00         .00           .00                     .00

COST CODE TOTAL                                       .00
*********************************************************************************************************************************

SUB. JOB ESTIMATE TOTALS    0.00        0.00         0.00                    .00

P.O. JOB ESTIMATE TOTALS    0.00        0.00         0.00                    .00

MAT. JOB ESTIMATE TOTALS    0.00        0.00         0.00                    .00

LAB. JOB ESTIMATE TOTALS    0.00        0.00         0.00                    .00


TOTAL JOB ESTIMATE FOR      0.00        0.00         0.00                    .00
SAMPLE JOB
```

Estimate listing.

Name _____ Date _____
Purpose of Trip _____
Week Ending _____
Job Name _____ Job No. _____

	SUN.	MON.	TUES.	WED.	THURS.	FRI.	SAT.	TOTAL	Cost Code
Date									
Destination									
Transportation									
Air									
Rental									
Parking & Tolls									
Limousine									
Breakfast									
Lunch									
Dinner									
Entertainment									
Room									
Telephone									
Tips									
Postage									
Laundry									
Miscellaneous									
Total									

Expense Report Form

Miles _____
@ 15¢ _____

GRAND TOTAL _____

Less Advance _____
Due Employee _____
Due Company _____

I certify this is an accurate record of my expenses:

_____   _____   _____
Signature   Date   Approved by

Expense report form.

JOB NAME SAMPLE JOB
JOB NO.   0000000

HISTORICAL DATA BASE   :UN 0/00/00   PAGE 00

C.C.'   C.C. DESCRIPTION   QUANTITY   U/M   PO-MAT $ SUBCONT $ PROD.   MAN HRS   AVG WG   MAN HRS $ TOTAL $

00000   SAMPLE COST CODE   -,---,---. --   -,---,---. -,---,---. -,---,---. -,---,---. -,---,---. -,---,---. -,---,---.

Historical data base.

INVOICE APPROVAL FORM

MO.-YR. (1)	VENDOR NO. (2)	P.O. NO. (3)	(4)	SUB-CONTRACT NO. (5)	(6)
JOB # (7)			MAT'L. (8)	SUB (9)	
DUE DATE	INVOICE #	COST CODE	QTY.	$ AMOUNT	CODE
(10)	(11)	(12)	(13)	(14)	(15)
APPROVED BY (16)					

157

```
MAINTENANCE UPDATES - ESTIMATES                              DATE RUN 0/00/00        PAGE    00
DIV. ____  PROJ.MGR. _____           JOB NO. & NAME 0000000    SAMPLE JOB     SYSTEM MAINTENANCE FORM NO. ____
**********************************************************************************************
           SYSTEM MAINTENANCE DATA ENTERED THIS FORM     *    CUMULATIVE ESTIMATE DATE INCLUDING THIS FORM
                                                         *
                                   EST.                  *                         EST.                    EST UNIT COST
COST CODE  DOC TYPE-# UM   *                             *                                                      OR
----------------------*  EST.QTY. EST.MAN HRS. TOT COST  PROJ. QTY.  *  EST.QTY. EST.MAN HRS.  TOT COST  PROJ QTY   AVG WAGE
COST CODE DESCRIPTION ***************************************************************************************
00000
SAMPLE COST CODE DESCRIPTION    .00       .00       .00       .00     *     .00        .00        .00        .00


NET MAINT.FORM ADJUST           .00       .00       .00       .00          ERRORS THIS FORM    00000
```

Maintenance updates—estimates.

MONTHLY TRANSACTION REPORT   DATE RUN 0/00/00   PAGE 00

VEND. #	VENDOR NAME	PO/SUB #	INVOICE #	JOB #	C.C. #	AMOUNT	INVOICED QUANTITY	U/M	SUB.	MATERIAL	P.O.
0000	SAMPLE	0000000	000			.00			.00	.00	.00
		COST CONTROL TOTAL									
		COST CONTROL TOTAL									

.00   TOTAL THIS JOB   .00

SAMPLE JOB

Monthly transaction report.

PROJECT CLOSE-OUT REPORT

Project Name:_____   Date Prepared:_____

Prepared By:_____   Job Number:_____

1. PROJECTED CONTRACT STATUS
   A. Projected Final Contract Amount $_____
   B. Projected Final Work-In-Process $_____
   C. Projected Final Overhead & Profit (Loss) $_____

2. ANALYSIS OF PROJECTED FINAL W.I.P.
   A. Work-In-Process $_____
   B. Accounts Payable (Itemize) (Use additional sheets if necessary)
      Vendor   Amount
      1._____ $_____
      2._____ $_____
      3._____ $_____
      4._____ $_____
      5._____ $_____
      6._____ $_____
      7._____ $_____
      8._____ $_____
      9._____ $_____
      10._____ $_____
      11._____ $_____
      12._____ $_____
                      Sub-Total Accounts Payable $_____ $_____
   C. Credits or Backcharges
      1._____ $_____
      2._____ $_____
      3._____ $_____
      4._____ $_____
      5._____ $_____
                              Sub-Total Credits $_____ $_____
         TOTAL PROJECTED W.I.P. (Must Equal 1.B. Above) $_____

3. WARRANTY ESTIMATES
   A. Labor $_____
   B. Material $_____
   C. Legal (By President Only) $_____

Project close—out report.

PROJECT STATUS REPORT # _____   DATE: _____
                                     JOB NO: _____

PROJECT: _____
PROJECT MANAGER: _____ SUPERINTENDENT: _____

PROJECT STRUCTURE	TOTALS	ANTICIPATED COST	ANTICIPATED PROFIT	PROFIT/COST %
1. Original Contract				
2. Change Orders (Thru C.O.# __)				
3. Sub-Total				
4. P & L Forecast				
5. Total				

BILLING STRUCTURE

6. Total Contract Billed _____   % Complete: Line $\frac{8}{5}$ = _____ %
7. Less Contract Retainage _____   (Cost)
8. Less Work In Process _____   % Complete: Line $\frac{12}{5}$ = _____ %
9. Plus W.I.P. Retainage _____   (Cost)
10. Total Overdraw _____   % Overdraw: Line $\frac{10}{8}$ = _____ %
11. Volume of Subs Bonded _____
12. W.I.P. (Less stored mat'ls.) _____   % Time Used: Line $\frac{19}{21}$ = _____ %
13. Interest This Month _____
14. Interest Cumulative _____

BUDGET ANALYSIS	ANTICIPATED COSTS	W.I.P.	%
15. Labor			
16. SC - PO - MAT(Less equip.)			
17. Equipment			

COMPLETION STATUS	DATE	DAYS	OWNER'S REPRESENTATIVE STATUS	
18. Original Contract			Excellent Relationship	☐
19. Elapsed Time			Cooperates	☐
20. Original Completion			Questionable Attitude	☐
21. Extended Completion			Difficult Relationship	☐
22. Estimated Completion				

PROJECT MANAGER'S REMARKS   W.I.P.-Mo.$ _____ Requested C/O$ _____ Days _____

_____
_____

DIVISION MANAGER REMARKS _____
_____

Due Second Working Day After Receipt of Monthly Job Summary

Project status report.

PURCHASE ORDER

SHOW THIS INFORMATION ON ALL INVOICES

PURCHASE ORDER NUMBER: № 3401

PROJECT NUMBER:

COST CODE (1) _____ COST CODE (2) _____

TO:

PROJECT:
DATE:
TERMS:
SHIP TO:
C/O _____
ADDRESS _____
DATE OF DELIVERY _____

F.O.B. _____

_____ COPIES OF SHOP DRAWINGS TO BE IN THE CONTRACTORS HANDS NO LATER THAN

SPECIAL INSTRUCTIONS:
1. INVOICES MUST BE IN DUPLICATE
2. THE ATTACHED ACKNOWLEDGEMENT COPY MUST BE SIGNED AND RETURNED WITHIN 10 DAYS. THIS ORDER IS SUBJECT TO CANCELLATION IF SIGNED COPY IS NOT RETURNED WITHIN 10 DAYS.
3. SHIPMENT MADE AGAINST THIS PURCHASE ORDER IN THE ABSENCE OF A SIGNED ACKNOWLEDGEMENT WILL CONSTITUTE SELLERS ACCEPTANCE OF ALL TERMS AND CONDITIONS HEREOF.
4. ALL FREIGHT SHALL BE PREPAID. IF FREIGHT CHARGES ARE FOR CONTRACTORS ACCOUNT, ADD TO THE INVOICE.
5. TAKE SPECIAL NOTE OF ATTACHED CLAUSES WHICH ARE A PART OF THIS ORDER.

SELLER AGREES TO DELIVER _____ WEEKS AFTER RECEIPT OF APPROVED SHOP DRAWINGS BY SELLER OR ITS AGENT.

VIA _____

RECEIPT AND ACCEPTANCE OF THIS ORDER, INCLUDING TERMS AND CONDITIONS ON REVERSE SIDE HEREOF, IS HEREBY ACKNOWLEDGED.

FIRM _____
BY _____
DATE _____

BY _____
TITLE _____

ORIGINAL TO VENDOR   (SEE REVERSE SIDE FOR TERMS AND CONDITIONS)   NO _____

Purchase order.

FROM: _____

_____

PROJECT:
PAYMENT REQUEST NO. _____
PERIOD _____ , 19___ , to _____ , 19___ .

STATEMENT OF CONTRACT ACCOUNT:
1. Original Contract Amount $ _____
2. Approved Changes (Net) (Add/Deduct) (As per attached breakdown) $ _____
3. Adjusted Contract Amount $ _____

4. Value of Work Completed to Date: (As per attached breakdown) $ _____
5. Value of Approved Change Orders Completed to Date: $ _____
    (As per attached breakdown)
6. Materials Stored on Site: (As per attached breakdown) $ _____
7. Total (4 + 5 + 6) $ _____
8. Less Amount Retained (_____ %) ($ _____)
9. Total Less Retainage $ _____
10. Less Previous Payments ($ _____)
11. AMOUNT OF THIS REQUEST $ _____

CERTIFICATE OF THE SUBCONTRACTOR:

    I hereby certify that the work performed and the materials supplied to date, as shown on the above represent the actual value of accomplishment under the terms of the Contract (and all authorized changes thereto) between the undersigned and _____ relating to the above referenced project.

    I also certify that all laborers, materialmen, suppliers, contractors, and subcontractors used on or in connection with the performance of this contract have been paid in full, except as noted on reverse side. I further certify I have complied with all Federal, State and local tax laws, including Social Security laws and Unemployment Compensation laws and Workers' Compensation laws insofar as applicable to the performance of this Contract.

    Furthermore, in consideration of the payments received, and upon receipt of the amount of this request, the undersigned does hereby waive, release and relinquish any and all claims under any applicable surety bond, rights of lien upon the above premises, and causes of action which the undersigned may now have or hereafter acquire, including, but not limited to those rights as contemplated by Chapters 255 and 713, Florida Statutes, except for rights to the extent that payment is retained pursuant to written agreement or payment to become due for work performed subsequent to the date hereof.

Date _____

                                                        SUBCONTRACTOR

Subscribed and sworn before me this _____ day of
_____ , 19 _____ BY: _____
                                                       (authorized signature)
Notary Public: _____
My Commission Expires: _____ TITLE: _____

TWO PART: WHITE COPY — MAIL TO   YELLOW COPY — SUBCONTRACTOR'S COPY

Subcontractor's application for payment.

SUB-CONTRACT AGREEMENT

SHOW THIS INFORMATION ON ALL INVOICES

SUBCONTRACT NUMBER: N° 1429
PROJECT NUMBER:
COST CODE (1) _____ COST CODE (2) _____

Contractor:

Sub-Contractor:

Project:

Owner:

Architect/Engineer:

Sub-Contract Date:

Sub-Contract Price:

Progress payment invoices in duplicate due in Contractor's office .... day of each month to be paid as conditioned below on .... day of each/next month.

The attached acknowledgment copy must be signed and returned within ten (10) days of the above Sub-Contract date. This agreement is subject to cancellation if signed copy is not returned within ten (10) days of the above Sub-Contract date. Work performed under this Sub-Contract agreement in the absence of a signed acknowledgment will constitute Sub-Contractors acceptance of all terms and conditions hereof.

The above terms are incorporated into and are more fully explained in the provisions that follow:

For the consideration hereinafter named, the Sub-Contractor covenants and agrees with the Contractor, as follows:

First: (A) The Sub-Contractor agrees to provide everything required to complete and will complete the following described items of work in connection with the construction of the project and will furnish all labor, materials, scaffolding, equipment, machinery, tools, apparatus, transportation, all required shop drawings, all required samples, and shall, as often as directed by the Contractor, completely clean all work and remove all debris from the job site, and perform all work necessary to complete ....

....

and as shown and called for on the plans and described in the specifications defined herein. Sub-Contractor hereby acknowledges that he has read, received, and is familiar with said plans and specifications.

(B) All of the provisions of the Prime Contract between Contractor and Owner are to govern the work under this Sub-Contract and Sub-Contractor is bound by all of the terms of the Prime Contract.

(C) Sub-Contractor shall submit ...... sets of required shop drawings and/or submittal data to Contractor no later than ....

(D) The Sub-Contractor is to secure, pay for, and file with the contractor, prior to commencing work, all certificates for Workmen's Compensation, Public Liability, Property Damage Liability Insurance, and such other insurance coverages as may be required by the specifications and addenda thereto, in the following amounts:

Bodily Injury .... Per Person   Property Damage .... Each Accident
                .... Each Accident   .... Aggregate
                .... Aggregate

(E) Sub-Contractor shall give bond payable to Contractor in the sum of .... on a form and with a surety satisfactory to Contractor for the faithful performance and payment of this Sub-Contract.

(F) The Contractor shall pay Sub-Contractor for work performed hereunder by voucher upon receipt of payment from the owner, an amount equal to .... percentage (...................... %) of the completed work performed from month to month, provided Sub-Contractor has delivered an estimate of the work performed during each month for which payment is sought. The monthly progress payment is expressly conditional upon the completed work complying with specifications as certified by the Owner or the Owner's Architect, Engineer or other representative as may be specified in the prime contract and is subject to the approval of the Architect or Engineer's judgement that the unpaid balance of this Sub-Contract shall at all times be sufficient to complete the Sub-Contract work required hereunder and to satisfy any unpaid liens or claim for which Contractor is responsible hereunder. Sub-Contractor will not remove any material stored on the job site for which payment is requested. Sub-Contractor agrees that Contractor shall not be liable for or responsible to Sub-Contractor in the event any property, materials, equipment, tools or other articles owned or used by Sub-Contractor are lost, damaged, removed, or stolen from the project site. Sub-Contractor shall submit to Contractor satisfactory evidence of payment of all indebtedness incurred for material and labor under this contract. The final payment shall be made .... days after completion of the work, upon issuance of certificate from the Architect or Engineer that the work has been done to his satisfaction, provided the Contractor has received payment for work performed hereunder from the Owner. This certificate shall be condition precedent to the right of Sub-Contractor to final payment. An additional condition precedent to final payment shall be the furnishing of a final and complete release of lien that the Sub-Contractor has paid all labor, material and suppliers and others, that could claim a lien by reason of Sub-Contractor's employment. Sub-Contractor shall furnish to Contractor all as-builts and/or construction manuals together with any other documents required by all contract documents referred to herein and the furnishing of the same shall be an additional condition precedent to Sub-Contractor receiving final payment. If at any time there shall be evidence of any lien or claim for which Contractor or Owner might become liable and which is chargeable to Sub-Contractor, the Contractor shall have the right to retain out of any payment due or to become due, an amount sufficient to indemnify Contractor and Owner against such lien or claim and charge or deduct all the cost of defense including interest, attorney's fees, court costs, appellate attorney's fees, and appellate court cost and payments(s) thereof if directed by court order or judgment. Should any claim or liens develop after all payments are made, Sub-Contractor shall refund to Contractor all monies that the latter may be compelled to pay in discharging such claim or liens, including interest, attorney's fees, court costs, appellate attorney's fees, and appellate court costs incurred by the Contractor in satisfying such claim or liens, or incurred in collecting said monies from the Sub-Contractor.

TAKE NOTE OF ATTACHED CLAUSES ON REVERSE SIDE WHICH ARE A PART OF THIS CONTRACT

COPY DISTRIBUTION: (ORIGINAL—WHITE) (ACKNOWLEDGEMENT—YELLOW) (PROJECT MANAGER—GREEN) (ACCOUNTING—PINK)

Sub—contract agreement.

```
DIV -                                          MONTHLY LABOR SUMMARY                                PAGE 00
                            PROJ MGR-        JOB NO. & NAME 0000000  SAMPLE JOB
                            IN PLACE QTY THRU 0/00/00    PAYROLL THRU 0/00/00    MAINTENANCE THRU FORM NO.  00
*************************************************************************************************************
CODE CODE / COST CODE DESCRIPTION            /UM                *                                          *
---------------------------------------------------             *           PROJECTED              PROJECTED*
                                                         PROJ.  PROJ.   PROJECTED   PROJ.          PROJECTED*
        QTY.  AVG.   PROD.  PROD.  * TOTAL     TOTAL     FINAL  AVG.    PROD.  PROD.  TOTAL         FINAL   *
              WAGE   U/MH   MH/U   MAN HRS     COST      QTY    WAGE    U/MH   MH/U   MAN HRS.      COST    *  VARIANCE
                                                                                                           *
*************************************************************************************************************
  00000       SAMPLE COST CODE                        *                                                    *
EST.     0     .000   .0000  .0000    LS       .00    *                                                    *
WKLY     0     .000   .0000  .0000     0       .00    *   0    .000    .0000    0      .0000    0    .00   *    .00
CUM      0     .000   .0000  .0000     0       .00    *
EST.BALANCE                                            *
*************************************************************************************************************

J O B   T O T A L S      QTY.          MAN HRS         COST      BALANCE    PROJ FINAL QTY    PROJ MAN HRS    PROJ COST    COST VARIANCE
             EST.         .00            .00            .00         .00
             WKLY         .00            .00            .00                        .00              .00           .00           .00
             CUM          .00            .00            .00
```

Monthly labor summary.

Check
REIMBURSEMENT SUMMARY

JOB NAME: _____ REQUESTED BY: _____
DATE: _____ APPROVED BY: _____

DATE	CHECK		VENDOR NAME	COST CONTROL ALLOCATION		TRANSACTION DESCRIPTION
	NUMBER	AMOUNT		NUMBER	AMOUNT	
		TOTAL:			TOTAL:	

Check reimbursement summary.

PETTY CASH
REIMBURSEMENT SUMMARY

PETTY CASH CUSTODIAN: _____   DATE: _____

JOB # / ACCT. MAJOR (p)	WORK IN PROCESS EXPENDITURES		
	COST CODE (p)	AMOUNT (p)	JOB NAME

REQUESTED BY: _____   APPROVED BY: _____

Petty cash reimbursement summary.

WEEKLY LABOR SUMMARY						WEEK ENDING 0/00/00 000000		PAGE 00
DIV.		PROJ.MGR.			JOB # & NAME			
COST CODE	COST CODE DESCRIPTION	UM	QUANTITY	AVG.WAGE	PROD. U/MH	PROD. MH/U	TOT MAN HRS	TOTAL COST
00		LS	.00	.000	.0000	.0000	.00	.00
00		LS	.00	.000	.0000	.0000	.00	.00
JOB TOTALS			.00				.00	.00

Weekly labor summary.

```
DIV -                         PROJ MGR-              SUBCONTRACT, P.O. & MATERIAL SUMMARY FOR    0/00/00                                      PAGE 00
                                                                    JOB NO & NAME  0000000   SAMPLE JOB
                                                     PAYABLES THRU  0/00/00         MAINTENANCE THRU FORM NO.   000
***************************************************************************************************************************************************
 COST CODE / COST CODE DESCRIPTION / UM  *                                           *                                *
 --------------------------------------  *         A C T U A L                       *       P R O J E C T E D        *
       E S T I M A T E D                 *                                           *                                *
  QTY    UNIT COST        COST           *   QTY   UNIT COST        COST     RETAIN. *  QTY    UNIT COST       COST   *  VARIANCE
***************************************************************************************************************************************************
  00000     SAMPLE COST CODE
 SUB.-1   0     .0000              .00         0      .0000            .00      .00        0      .0000           .00         .00
 MO.                 SUBCONTRACTOR'S NAME      0      .0000            .00      .00
 CUM                 SUBCONTRACTOR'S NAME      0      .0000            .00      .00
 EST.BALANCE

 SUB.-2
 MO.                 SUBCONTRACTOR'S NAME
 CUM                 SUBCONTRACTOR'S NAME
 EST.BALANCE

 P.O.-1
 MO.                 VENDOR'S NAME
 CUM
 EST.BALANCE

 MAT.
 MO.
 CUM
 EST.BALANCE

 EST.TOTAL COST                     .00                ACTUAL TOTAL COST       .00
***************************************************************************************************************************************************

 TOTALS FOR JOB NO & NAME  0000000    SAMPLE JOB
 SUB    0                                       0                    .00                          0                   .00         .00
 P.O.
 CUM                                .00         0                    .00                          0                   .00         .00
 MAT
 CUM                                .00         0                    .00                          0                   .00         .00

 TOTAL WIP                          .00
 EQUIPMENT CODES                   (.00)                                              INTEREST INCOME
 TOTAL WIP LESS EQUIP.              .00
```

Subcontract, P.O. and material summary.

TIME CARD

JOB # ________ JOB NAME ________________________________ WEEK END ____________

EMP # ________ EMP NAME ________________________________ CRAFT CODE ____________

DATE →																	
DAY OF WEEK →																	
C.C. #	ALT. CRAFT CODE	TOTAL STRAIGHT	TOTAL OT	STRAIGHT	OT	STRAIGHT	OT	STRAIGHT	OT	STRAIGHT	OT	STRAIGHT	OT	STRAIGHT	OT	STRAIGHT	OT
TOTAL																	

Time card.

UNIT COMPLETION REPORT

JOB #0000   JOB NAME   SAMPLE JOB   WEEK ENDING 0/00/00

```
*********************************************************************
*             *       *           * PREVIOUS  *          *           *
* COST        *       *ESTIMATED  * TOTAL QTY *QUANTITY  * TOTAL     *
* CODE        *  *UM  *QUANTITY   * AS OF     *COMPLETE  *QUANTITY   *
*   #         *       *           * 0/00/00   *THIS WK   *COMPLETE   *
*             *       *           *           *          *           *
*********************************************************************
*             *       *           *           *          *           *
* 00000  SAMPLE COST CODE  *      *    0.00   *   0.00   *   0.00    *   0.00
*             *       *           *           *          *           *
---------------------------------------------------------------------
```

Unit completion report.

KEY WORD INDEX

A

Accounts payable, 12, 65, 140
Accounts receivable, 65, 140
Actual cost, 23

B

Backcharges, 57, 140

C

Change order, 44, 140
Cost:
   actual, 23, 140
   estimated, 23, 140
   projected, 140
Cost Accounting System, 19, 37, 43, 140
Cost code, 14, 33, 36, 53, 57, 60, 140
Cost forecasting, 140

D

Delivery ticket, 140

E

Estimate, 140
Estimate listing, 1, 6, 25, 140

G

General conditions, 142

H

Historical data, 9, 66, 142

I

In-place quantity, 18, 20
Invoice, 14
   payable, 142
Invoice approval form, 12, 13

L

Labor, 4, 41, 68, 142
Labor hour, 142
Labor tax, 6, 75, 142

M

Maintenance form, 143
Material, 12, 25, 41, 143
Monthly Job Summary Report, 23, 145

P

Project cost, 23
Project Status Report, 23, 143
Purchase order, 7, 10, 12, 25, 41, 144

R

Risk management, 144

S

Source Document, 2, 8, 27, 144
Subcontract, 7, 12, 24, 41, 144

Purchase Order and Materials Summary, 24, 34, 145
Summary:
  monthly job, 23, 145
  monthly labor, 24, 30, 145
  weekly labor, 22, 145

U

Unit Completion Report, 145

V

Variance Reporting Procedures, 4, 9
Vendor, 15, 146

W

Weekly time cards, 19
Work in process, 23, 40, 146